Helmut Mothes

Process Design – Synthesis, Intensification, and Integration
of Chemical Processes

Series:

Innovation in Manufacturing Technologies

Process Design
Synthesis, Intensification, and Integration of Chemical Processes

Helmut Mothes

ISBN 978-3-00-049149-8

Bibliographic information published by the Deutsche Nationalbibliothek
The Deutsche Nationalbibliothek lists this publication in the Nationalbibliografie;
Detailed bibliographic data are available on the Internet at http://dnb.d-nb.de

www.manufective.com

Content:

Foreword

In the past process design aimed at solutions converting raw materials into products in a technically convincing and economically profitable manner. This effort has certainly created impressive, technical, and economic progress.

Today the world faces new challenges. Food supply for a growing world population, healthcare for an aging society, and global warming represent an on-going challenge to society. At the same time, biotechnology, nanotechnology, green technologies, and the internet of things offer new opportunities.

Future process design still requires sound chemical engineering skills and system thinking capabilities, but also an open mind-set and a serious commitment for efficient, sustainable "no-regret"-solutions in a complex, volatile future.

Process design can build on many excellent chemical engineering textbooks addressing unit operations. A few books cover the area of process synthesis, intensification, and integration.

This textbook communicates basic ideas about process design to chemical engineering students and encourages experienced engineers to reflect on – and perhaps challenge – their daily approach to process design.

Concepts, methodologies, and philosophies constitute the main topic of this book promoting a holistic process design methodology that introduces the concept of "no-regret"- solutions to the classic synthesis approach for chemical processes. These "no-regret"- solutions are process designs that give the priority to design solutions offering a robust and sustainable performance in all feasible future scenarios.

The book is a summary of my courses "Process Intensification" and "Process Design" given at the Technical University Dresden (TU Dresden - 2008), East China University of Science and Technology (ECUST Shanghai – 2012-2014) and Ruhr University Bochum (RUB – 2014-2015).

Prof. Lange encouraged me to present a 2-day lecture on process intensification at the TU Dresden. Prof. Weiling Luan gave me the opportunity to give a course on process design as guest professor at the East China University of Science and Technology (ECUST) in Shanghai. In Germany, Prof. Grünewald became my hoist and partner for a lecture on process design at the Ruhr University Bochum (RUB).

To a large extend, my book on process design builds on experiences and materials from Bayer Technology Services. The book would not be feasible without the input of many Bayer people – technical contributions, exciting suggestions, and enlightening discussions.

1 Introduction

The chemical industry represents a mature business. During the last 150 years, the production facilities and supply chains of the chemical and pharmaceutical industry have developed into complex, global systems. Processes in the chemical industry have reached a sophisticated technological level.

In the past, the chemical industries could rely on steadily growing economies creating a stable and predictable framework for product demand, raw material availability, and technological progress. This framework provided a reliable and sound basis to design chemical processes, plants and whole supply chains.

Today an intensive, global competition characterizes the chemical business. Shorter product cycles, unreliable raw material supplies, and disruptive technology development increase the risk that a process design – optimal for a fixed set of operational parameters – cannot flexibly respond to product demand fluctuations or changing raw materials in the future. Ecological necessities aggravate the requirements for sustainable process designs.

The "conventional" process design methodology is particularly suited for precisely defined design problems with straightforward solutions. A design problem of the past could rely on predictable boundary conditions. This framework concerning resource availability and product demand combined with new sustainability goals add uncertainty to process design. Global production systems form a huge, multi-dimensional solution space for a process design task.

Complexity and *Uncertainty* evolve into key aspects of future process design.

New design methodologies require features to reduce complexity and manage uncertainty – from idea generation through process development and plant implementation.

Of course, a sound knowledge of chemical engineering remains essential for process design. In the future, a process methodology has not only to provide sound design algorithms for unit operations and equipment, but also to efficiently handle uncertainty and complexity. A design methodology must lead to process designs that react flexibly towards unexpected changes of the business environment.

This book discusses a holistic process design methodology dealing with uncertainty and complexity. The discussion on smart methodologies focuses on key design features and applies them towards selected design problems.

2   Process Design – A conceptual Approach

The ultimate task of a process design methodology is to create the blueprint of a chemical process converting raw materials into products. This blueprint is used in later design stages to engineer the process in detail and finally to construct a plant.

Figure 1 illustrates this task. A chemical process converts raw materials into an intermediate product such as plastic granules used to produce the consumer goods – from helmets to car parts and stadium roofs.

The design methodology should lead to efficient and sustainable process designs. Efficient processes convert raw materials into products with a minimum of material and energy input. Sustainable processes do not waste valuable resources and hurt

Figure 1: The Process Design Task (Source: Bayer, Leverkusen))

the environment. Yield, productivity or other engineering driven parameters primarily characterize the efficiency of a chemical process design. Ultimately, these scientific parameters are translated into economic parameters – particularly profitability.

Normally the product quantity and quality provide the input information. Process design defines the process structure and parameters to achieve the production goal (Figure 2).

During the past 50 years, many methodologies have been proposed and used to design chemical processes. Generally, design methodologies are tailored towards a specific task: grass root design, retrofitting, or optimization.

Before a holistic design methodology is illustrated, some general aspects of process design are discussed to provide some beneficial insights into process design.

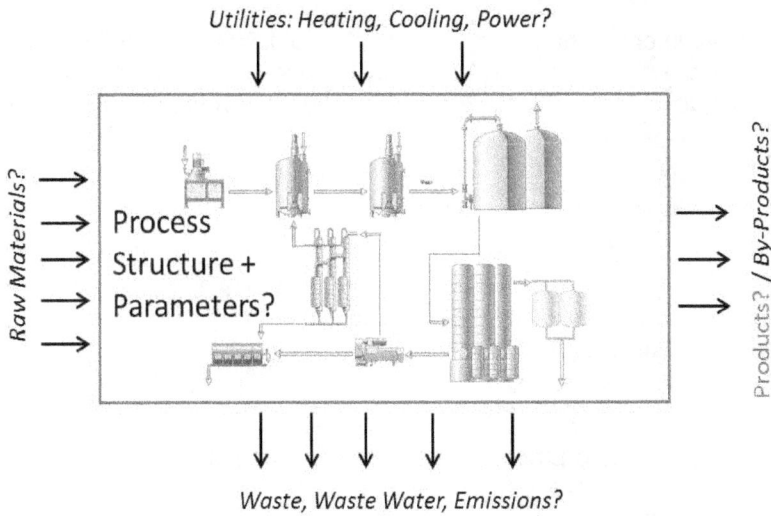

Figure 2: Input - Output Scheme for Process Design

Process design is part of a large effort to convert a process idea into a functioning chemical plant including process development, process design, basic engineering, detailed engineering, and plant construction.

Figure 3 gives a schematic description of the overall process leading to a new plant. Process development provides the input for process design. Process design translates this information into a process structure. This process structure often

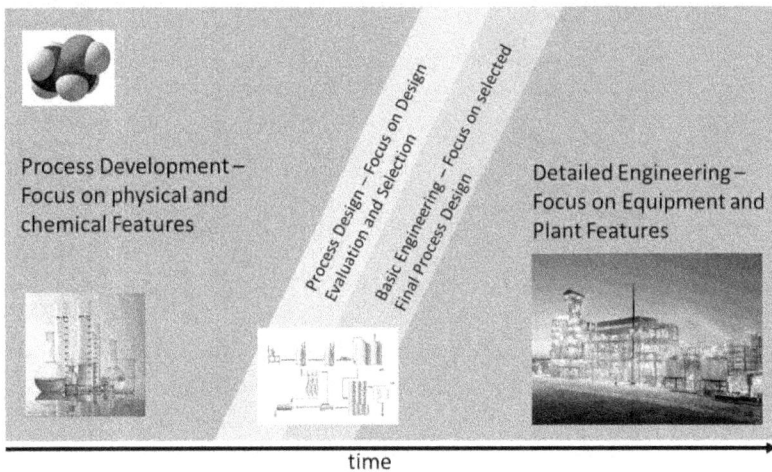

Figure 3: Phase Diagram for Process Design

remains unchanged during later project phases.

Process development contributes significantly to process understanding, while engineering phases supply information on process details such as equipment, piping, and other plant specifications. Both phases add a huge amount of detailed knowledge about the process and plant.

Figure 4 illustrates the impact of the respective project phase on process understanding, project cost, and profitability.

Process development and detailed engineering generate a lot of information about chemistry, property data, equipment, and process. They significantly contribute facts to understand the process. This fact generation requires resources and time resulting in significant cost. Process design – basic engineering to a lesser degree – converts data from process development into a final, complete process structure. This step only requires limited resources and cost.

Process development and process design define process technologies and select process structures that are further developed and optimized during basic and detailed engineering. In case process development and process design decide in favor of an inferior technology and process structure, a perfect optimization during basic and detailed engineering still leads to a sub-optimal solution with limited project profitability.

Excellent, basic, and detailed engineering are indispensable to create an operable, efficient, and profitable process – they are necessary, but not sufficient steps. Selections made during process development strongly impact on profitability. Process design plays a key role questioning the process technologies and

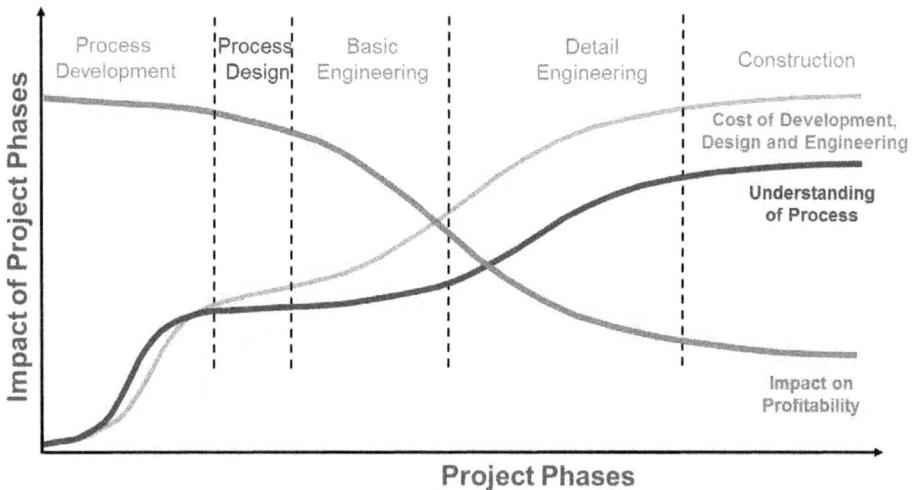

Figure 4: Impact of Design Phases on Design Results

providing the best process structure for basic engineering and detailed engineering.

In many industrial companies, basic engineering is part of the engineering department, while process design belongs to the process development organization. Both groups need to communicate intensively. Process designers and basic engineers must act as key interfaces. Process designers guide process development towards a realistic, economic process proposal, while basic engineers provide the optimal input for a successful detailed engineering.

Information gathering in process development and detailed engineering is linked to high cost and expenses, while process design and basic engineering lead to optimal selections evaluating already existing information. The impact on process profitability is high, while expenses are rather limited. Obviously, process design represents a key element to create economic and sustainable processes.

2.1 Design Framework – Complexity and Uncertainty

In the past, the main objective of process design was to find a technical solution to convert raw materials efficiently into products within a stable, predictable environment. Today process design occurs in a complex and volatile environment.

Intensive global competition, powerful consumer trends and fast emerging, disruptive innovations create a volatile environment. Process designs have to perform in a steadily changing environment endangering the fixed design framework of the past. Product demand is literally impossible to forecast reliably. Unstable economic and/or financial systems will not facilitate the problem.

A process design solution is no longer satisfactory operating successfully for a given set of design parameters. Process designs have to perform under variable conditions – fluctuating product demand, changing feedstock and strong price competition.

A comprehensive understanding of technical and societal trends facilitates the design of processes within a volatile environment. The short-term and long-term impact of these trends on society and environment adds a significant challenge to the design task.

Process design needs to ask questions about the economic, ecological, and societal necessities of the future:

- Which challenges do future trends generate for a process design?
- What does sustainability require for a process design?
- Which options do technological innovations offer to a process design?

Trends

Since we are mainly interested in insights describing the impact of future trends on the industrial production system, within which the new process has to perform efficiently, it is beneficial to distinguish between:

- Consumer megatrends

- Industrial productivity cycles
- Emerging enabling technologies

Consumer trends particularly affect the market pull (product quantity, quality, and differentiation), while industry productivity cycles characterize technological factors. Emerging technologies act as technological enablers.

A thorough understanding of the impact of consumer megatrends, productivity cycles, and emerging, enabling technologies on industrial systems is a prerequisite to anticipate future requirements for process designs successfully.

Which megatrends are likely to influence the environment within which a new process design has to operate in the future?

Globalization creates a worldwide competition of industrial systems and processes for market shares, novel technologies, and resources (materials and labor) as well. In the past, forecast of product demand was comparably easy due to steady economic growth. If the actual demand in one region deviated from the forecast, sales in other regions compensated this shortfall. In the future, a forecast of product demand becomes *uncertain* due to an intensive global competition.

Individualization leads to more tailor-made consumer products. Mobile phones, leisure articles, food, cars etc. differ in form, colors, and qualities (e.g. configurations). Individualization results in personalized drugs for healthcare and target-oriented products for agrochemical applications. Individualization results in more articles with smaller quantities making manufacturing and supply chains even more *complex*.

Sustainability represents another trend influencing the framework of a new process design. The selection of the best feedstock – oil, gas, or biomass for example – becomes increasingly more difficult.

The optimal design of a process, however, is not only affected by consumer trends, but also by technical trends. Kondratieff has first described *Industrial Productivity Cycles* (Händeler, 2011). The so-called Kondratieff cycles assume that technological innovations determine the development of societies. Innovation generates wealth in waves. Industrial productivity cycles are always linked to economic cycles.

The innovation of the steam engine replaced human power by machine power allowing production to move from labor based to steam driven manufacturing. This development changed whole industries such as the textile industry. The construction of railway systems revolutionized the transportation of people and goods, once more opening-up broad business opportunities. In a next cycle, electricity and science based chemistry allowed decentralized manufacturing of many new products – from consumer goods to drugs. Mass mobility with cars became the core of the next productivity cycle generating value in many areas. Information technology with computers created productivity gains in the last cycle.

There are different predictions on the next industrial production cycle. Händeler proposes a next cycle built around the future healthcare systems (Händeler, 2011). Others consider the internet of things as the next cycle theme (Anderson, 2013).

Brynjolfsson considers computer power, digitization, and automation as drivers of a *second machine age* (Brynjolfsson, 2013):

What are the likely innovation drivers of the next cycle?

Biotechnology and N*anotechnology* enable a more efficient synthesis of chemicals in a new process design. *Additive Manufacturing* is a novel, disruptive production technology. Progress through *Digitalization* and *Communication* makes efficient processes and supply chains possible, since increased computational power, a switch from analog to digital data processing, and communication through the internet result in a material management on a previously unknown performance level. Data mining, for example, creates unknown possibilities for manufacturing.

What is the result, if the designer of a new process or plant does not take advantage of these opportunities?

What happens to a newly built process using classical catalysts, if a competitor can produce the chemical with synthetic biology or nano-catalysts more economically? Does 3D-printing due to inexpensive equipment challenge the "economies of scale"-paradigm of the process industries? How does the internet of things influence the design of future processes?

The interactions of megatrends, productivity cycles, and enabling technologies with an industrial manufacturing system add uncertain, dynamic boundary conditions to

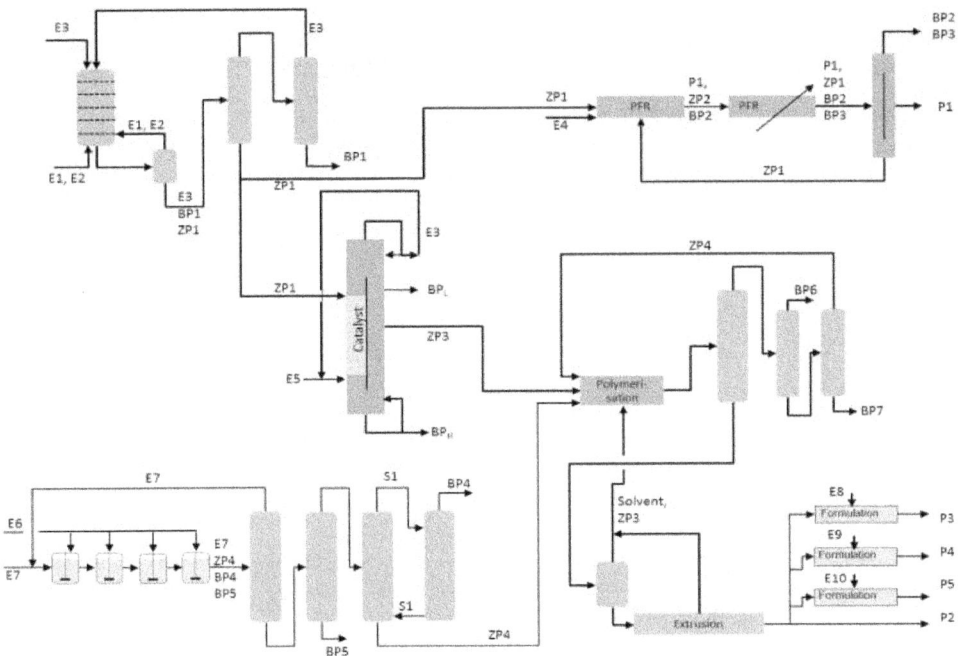

Figure 5: Process Design – The Process Network

process design those are difficult to be predicted in advance.

In addition to the volatile environment, finding an initial process structure is a "creative" step that complicates process design. Classical design strategy consisting of modeling, simulation, and optimization supported by experimental studies requires a process structure!

Figure 5 schematically describes a chemical production site with several linked processes. Obviously, chemical processes represent well-defined systems accessible to modeling and simulation, provided the process structure is available.

The search for the best process structure, however, is a qualitatively different task. Neither heuristic, hierarchical methods (artificial intelligence) nor mathematical algorithms (mixed integer non-linear programming) succeed in designing process structures – at least for realistic design tasks.

To create and evaluate all feasible structures does not represent an option to use modeling and simulation as a straightforward way to process synthesis, if the complexity is huge at least.

Why becomes process design generally a truly complex task?

In reality, the process that a process designer has to design is part of a process network described in Figure 5. On site, multiple interactions occur between various processes, utilities, logistics, and waste treatment. Products from one process become raw materials for other processes. Raw materials, intermediates, and products require sophisticated logistical systems. Utilities and heat recovery networks generate complex operational systems (Figure 6). Optimally designed individual processes do not necessarily lead to the site optimum.

Each production site is part of a global supply chain. This company supply chain competes with supply chains of other companies (Figure 7).

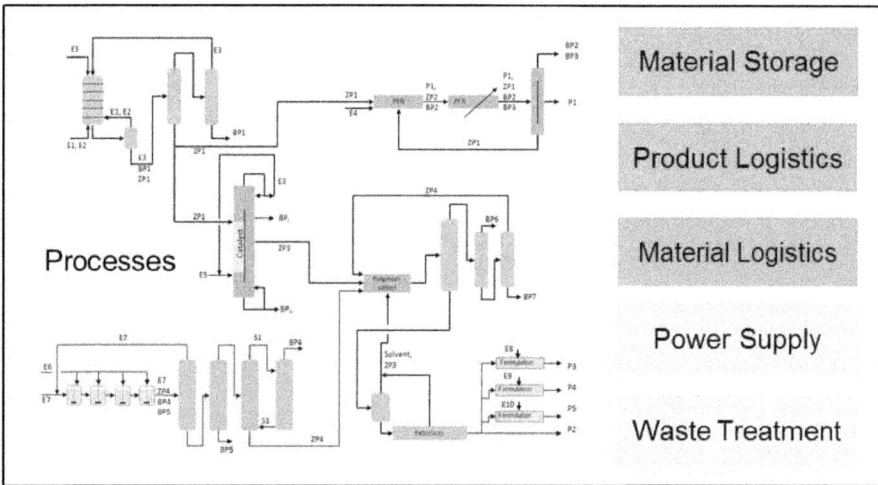

Figure 6: Process Design - Plant Site

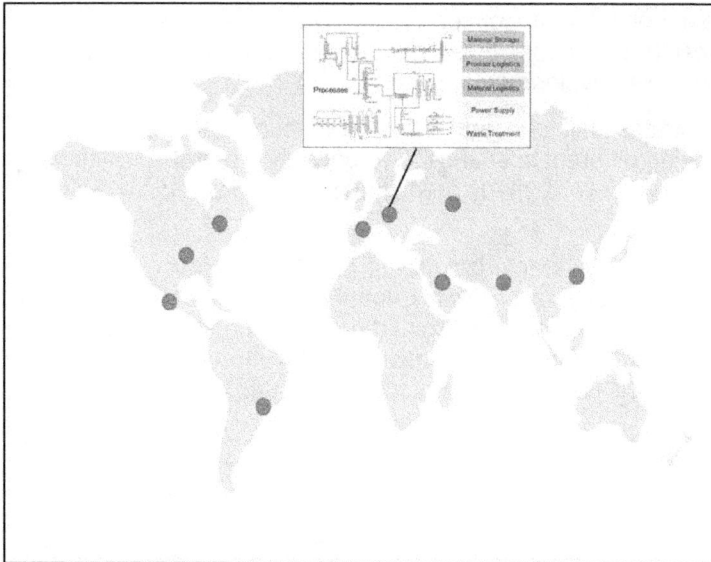

Figure 7: Process Design – The Supply Chain

The interaction between the process (to be designed) and the overall supply chain creates complexity unknown in the past when focusing on a single process task was sufficient. Adding the time dimension complicates the process design task even further.

Process design generally includes complex systems in an uncertain future of megatrends, industry cycles, and disruptive innovations. Although a prediction about the future is impossible, one statement is safe:

Uncertainty and complexity characterize the design framework for future processes, plants, and supply chains.

What are the methodological consequences, when traditional concepts may not work for complex systems embedded in an uncertain, technological, and cultural environment?

No-Regret Strategies

Process designers cannot focus on a single scenario in times of uncertainty and complexity. Instead of searching for the optimal process design for a probable, sometimes only desired future scenario, designers must take into account all feasible scenarios (Schwartz, 1991).

A multiple scenario strategy looks for a process design, which is not the optimal solution for a single, even highly probable scenario, but performs robust, flexible, and economically acceptable in all feasible future scenarios. There might be a better solution (higher productivity, more profit, etc.), but the strategy is not only to maximize the outcome, but also to minimize the likelihood for disaster, in case the future has not been described properly.

The rationale behind that concept is to overcome uncertainty with respect to simulated results by choosing the design that performs best with respect to all potentially feasible developments.

Figure 8 illustrates this concept. This figure shows the *future trend and event tunnel* containing different scenarios. Symbolically the graph outlines that the future will not be straightforward, many predictable and unpredictable events may change the course.

No-Regret Solutions characterize robust and flexible process designs performing acceptable in all feasible scenarios. Even if more economically advantageous, but also more risky solutions are available, there is an incentive for a no-regret solution to minimize the likelihood of a business disaster. In another words you select the solution that you will not regret in any future scenario.

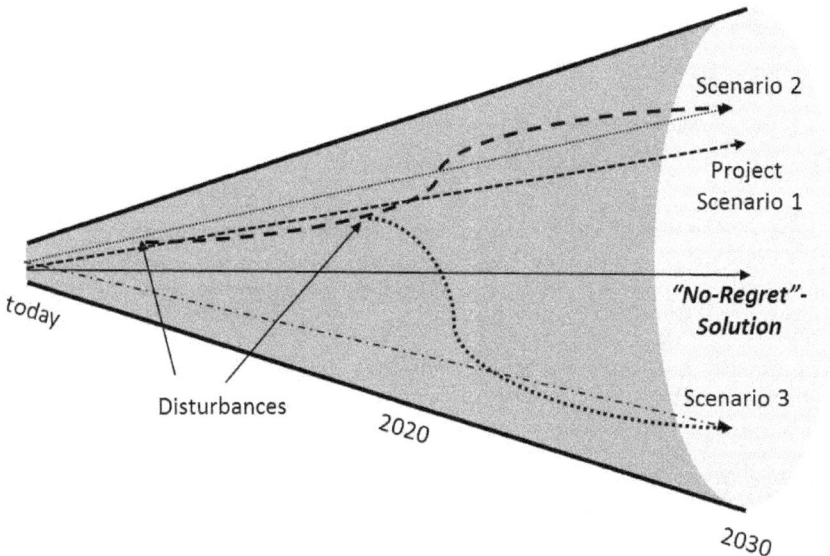

Figure 8: Process Design – The Future Scenario Channel

For example, the design of a new process to produce a chemical requires the selection of the best feedstock and respective process technologies. Oil, gas, or biomass is an alternate feedstock. Which process design is regarded a no-regret solution?

Today oil may appear the most economical feedstock based on current cost and availability. Gas, however, offers a superior process performance concerning carbon dioxide emissions. Biomass is preferable in terms of sustainability. Global competition for resources and effort for sustainability make a reliable forecast on the best feedstock a mission impossible.

The designer can either select process technologies tailor-made for oil, gas, or biomass or go for a technology processing multiple raw materials. This multi-purpose technology is likely to require higher investment compared to single-feedstock technologies.

Selecting a multi-purpose technology for the new process could represent a no-regret solution. Provided a multi-purpose technology meets a minimum profitability goal, the no-regret solution trades some profitability points against risk avoidance and more sustainability.

In a mature industry, process design often appears comparably easy, since the design task does not start from scratch, but builds on earlier designs, well-known technologies, and long-term, operational experience. The challenge, therefore, is to find the technical design, which performs optimally within future framework shaped by societal, economic, ecological trends.

No-regret solutions in process design combine profit maximization with risk minimization:

- *In process design, flexibility becomes an indispensable design criterion equivalent to efficiency.[1]*
- *Sustainable no-regret solutions require a new balance between profit maximization and risk minimization in times of uncertainty and complexity.*

---

[1] *Nevertheless, a sound knowledge of chemical engineering and other disciplines remains the indispensable foundation of process design.*

2.2 Design Strategy – A Journey into unknown Spaces

In general, design of a novel process is comparable with making an invention. Unfortunately, there is no recipe available, how to find an optimal process structure, since a "unified" theory or a "sudden" enlightenment of the designer is not feasible?

Since the designer has to start the creative design process somewhere, it is recommendable to begin the discovery journey at the core area of a chemical process (e.g. the reaction system) and then move step-by-step towards unknown regions (e.g. separation and formulation).

The traveler (e.g. process designer) carries in his backpack: a basic map (the chemical reaction system), some information on the landscape (material data), reports from earlier travelers (empirical data, simulation software) in the area and a survival guide (your methodology).

Two final comparisons between a traveler on his journey and a designer on his project address guides and backlashes. A traveler uses a compass or the stars to keep on track. A designer needs targets and benchmarks to control progress he makes. Traveler and designer may have to go back, if he cannot overcome obstacles.

This book focuses on the design methodology – particularly the method to create an initial process structure.

Are there general feature of the design methodology that can guide the designer? Do different design methodologies possess recognizable, common features?

Design Focus – Creating Structures

Process design generates an initial process structure linking unit operations.

Chemical processes always consist of three specific operations: reaction, separation, and formulation. The reaction section is necessary to convert feed materials into new chemical components by chemical or biochemical conversions. Separation and purification steps necessary to recovery the product in a pure form follow the reaction section. Subsequently, formulation converts this pure material in a material form that is appropriate for the next processing step or the final consumer good.

Recycle streams are a characteristic feature of chemical processes to utilize raw materials completely. Linking the unit operations of a process with material recycle flows completes process design. Figure 9 gives a systematic description of such a basic process design.

The unit operations for reactions consist of many alternate devices such as stirred tanks, tube reactors, loop reactors and many more. Many different unit operations are available for separation and formulation (Figure 9).

A chemical, pharmaceutical plant consists of a network of several reaction-, separation-, and formulation systems linked by material, energy, and information

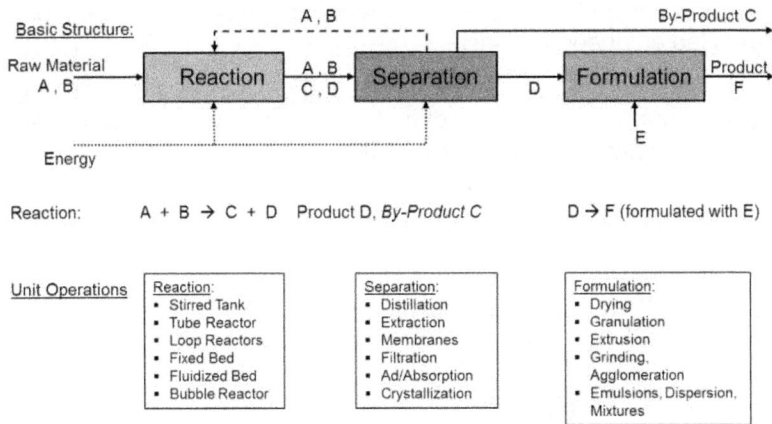

Figure 9: Process Synthesis Task

streams (Figure 10). The sequence of steps may change, but the basic elements remain the same.

A reaction system may include a sequence of different reactors. A separation system could generally consist of several sub-separations defined by separation sequence and separation unit technologies. The same holds true for formulation systems. The first system is not necessarily a reactions system, the final system is not mandatory a formulation system. Recycle streams generally add complexity to a process system.

Process design aims at defining the process structure, selecting the unit operations/equipment, and optimizing the process parameters. Figure 11 shows the result of a process design, a flow sheet containing the process structure, unit

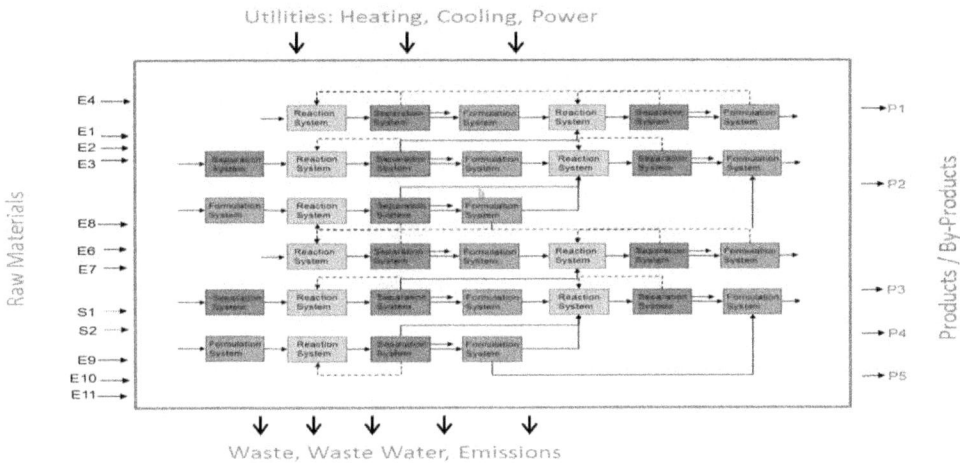

Figure 10: Network of Reactions, Separations and Formulations

operations/equipment and process parameters. This then serves as input for the basic and detailed engineering phases.

During process design, it is beneficial to recall that process design tries to generate the process structure and select the optimal process parameters. Obviously finding the proper process structure is more challenging than calculating the optimal

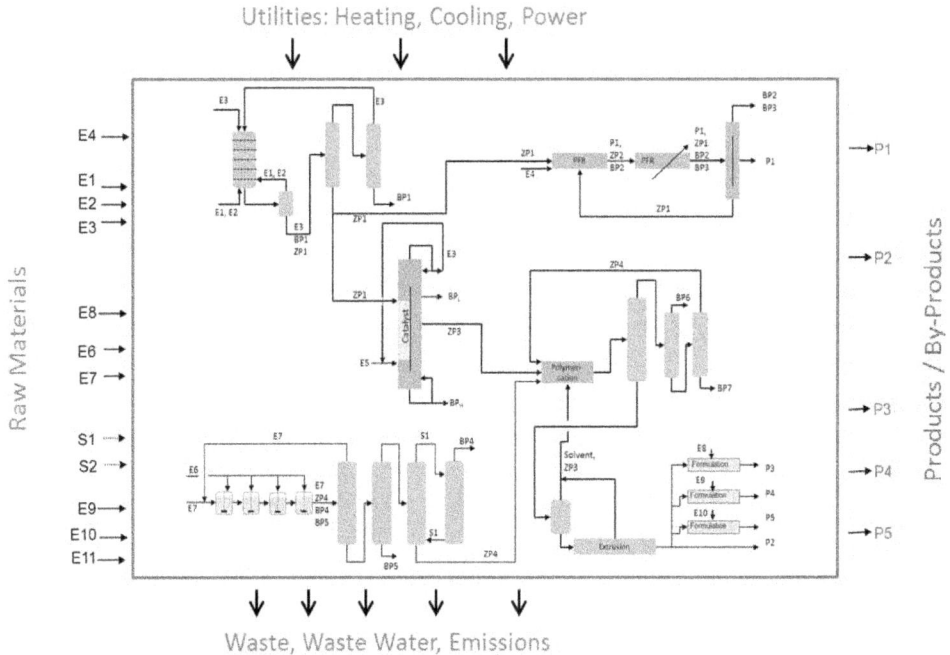

Figure 11: Process Design Result

parameters for a structure given. While process parameters can be defined using mathematical algorithms, the structural search faces essential mathematical obstacles.

Design Strategies

In the literature there are many excellent books on process design available (Douglas, 1988), (Seider, Seader, & Lewin, 2004), (Smith, 2004), (El-Halwagi, 2012). The methodologies outlined belong into two methodological categories:

- Hierarchical Concepts
- Superstructure Concepts

Either the designer adds element after element to create the final process design or establishes a process superstructure containing all feasible process elements subsequently removing inferior elements until the final process structure is accomplished. Figure 12 illustrates the two concepts. The left scheme outlines the

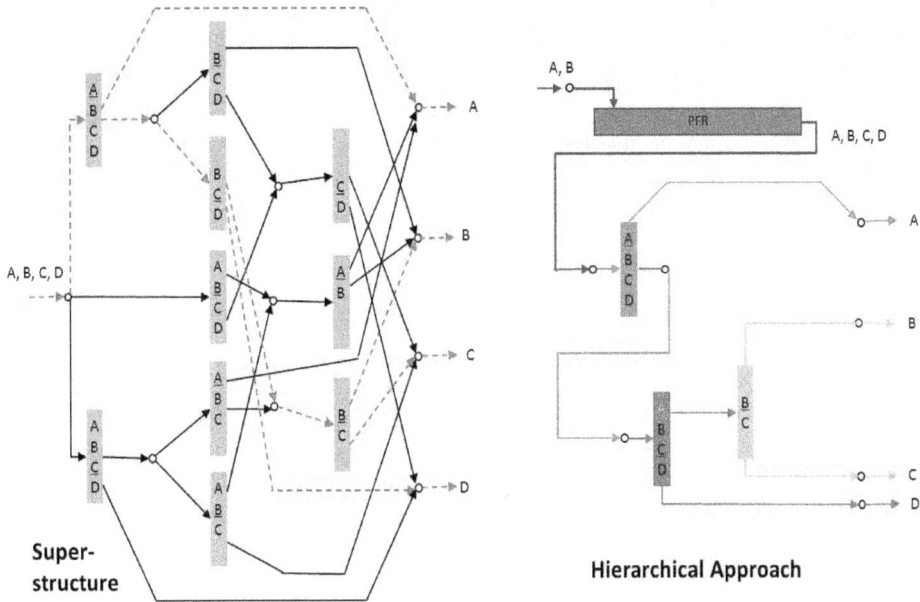

Figure 12: Different Process Design Methodologies

main element of the superstructure approach, while the right side demonstrates the hierarchical method.

Since the creative task to generate an optimal process structure is characterized by complexity due to many possible alternatives and a missing straightforward procedure, a guide facilitating the design procedure is indispensable.

The superstructure approach tries to overcome the creative challenge adding as many structures as possible in a superstructure that then should include the optimal solution. The hierarchical approach tries to reduce complexity by treating small sub-problems manageable one after the other.

The superstructure approach is likely to contain the optimal process structure, but its reduction to the optimal solutions becomes the bigger the more comprehensive the structure is set-up. The hierarchical approach is a manageable tool, but does not guarantee an optimal solution.

In reality, a designer' approach is a combination of both mental concepts.

The hierarchical Design Methodology

The hierarchical approach starts with the core operation of a process and adds design layer by layer to reach the final solution. The core operation is normally the reaction step subsequently adding recycle, separation/purification, and formulation. The challenge for this approach is to find the best criterion to make the right decision at each step. In addition, insights made at later steps may prove an earlier decision as wrong.

Figure 13 shows the hierarchical design approach. The design methodology follows the same three steps on each hierarchical level: find alternate structures, evaluate them, and select the best.

During the first step, the optimal reactor system is designed based on the reaction system information. Step 2 addresses the recycle structure, while level 4 defines separation sequence and separation technologies. A final evaluation of the process structure with respect to intensification of the unit operations and structure integration (systems engineering) occurs in step 3. [2]

To evaluate alternatives at each decision level represents the creative step of process design. Since a hierarchical approach divides the overall design problem

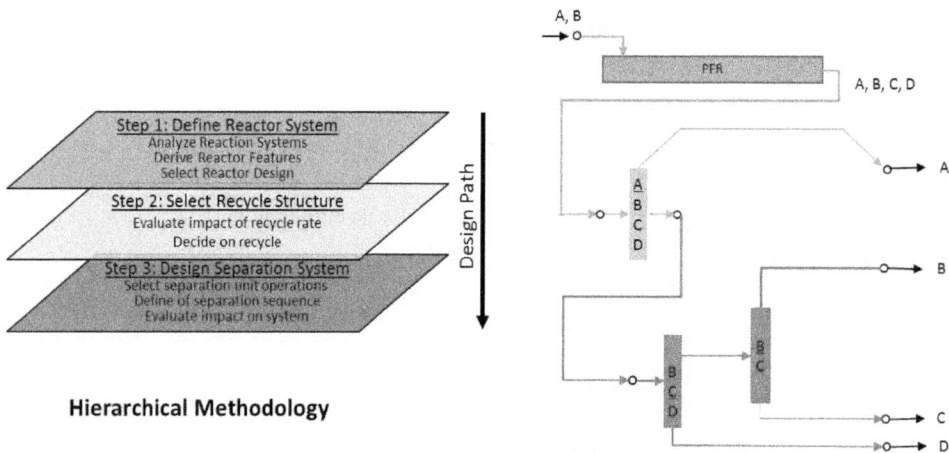

Hierarchical Methodology

Figure 13: The hierarchical Design Methodology

into manageable pieces, a complete evaluation of alternate designs becomes possible.

This mental model actually combines a hierarchical approach with a superstructure concept at each hierarchical level (Douglas, 1988), (Smith, 2004). Nevertheless, the methodologies cannot provide a fool-proven, cooking guidebook to establish the process structure.

The hierarchical approach faces a significant challenge.[3] If a decision on a later level leads with an earlier discarded structure to a better structure, this option would not be discovered, at all. For this reason, the designer requires criteria to decide on a superior solution at each level. The concept of economic potentials represents a promising tool to judge on alternatives (Douglas, 1988),

---

[2] Chapter 3.1 outlines a complete hierarchical approach with five levels.

[3] Douglas and Smith discuss the structural challenge in detail. The design approach of this book closely follows their ideas (Douglas, 1988), (Smith, 2004)

The Superstructure Design Methodology

The superstructure approach tries to overcome the creative dilemma to find an optimal structure by composing a complete superstructure – rather a sweat effort than enlightening. Provided the superstructure contains the optimal structure, a systematic removal of inferior elements could lead to the optimal process structure. Removal of inferior elements from a superstructure is likely an easier – less creative, rather laborious – effort.

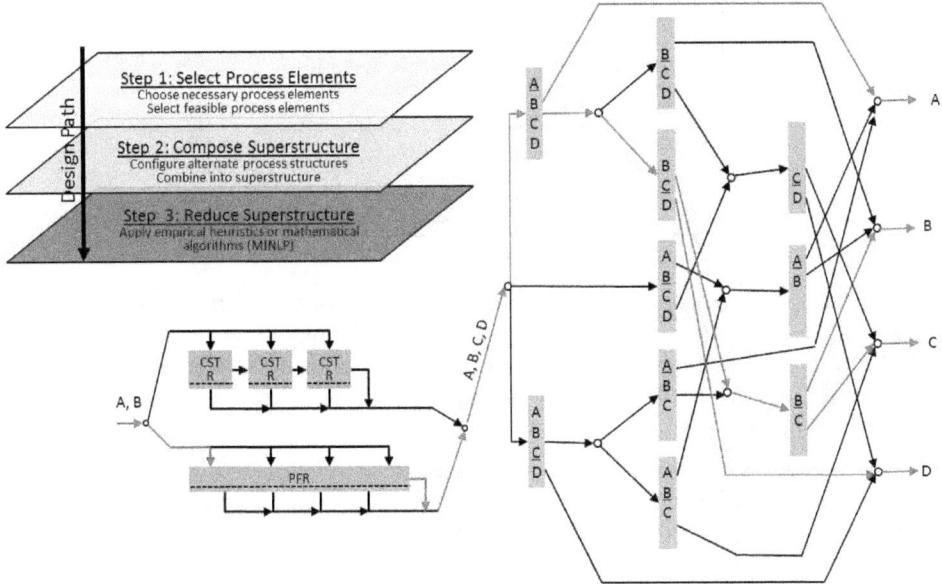

Figure 14: Superstructure Approach for Process Design

Figure 14 describes the three steps of a superstructure approach and illustrates a superstructure and the embedded optimal solution (dotted line). The procedure to derive the optimal solution involves three steps:

- Select suitable technologies
- Derive a superstructure (hopefully including the optimal structure)
- Remove inferior design elements generating an optimal structure

A simple reaction illustrates the superstructure concept.

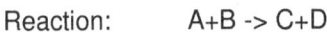

Reaction: A+B -> C+D

Material Data:

A, B, C, and D are liquids at ambient air pressure.

Boiling point: $T_A< T_B< T_C< T_D$

Separation of B/C is difficult.

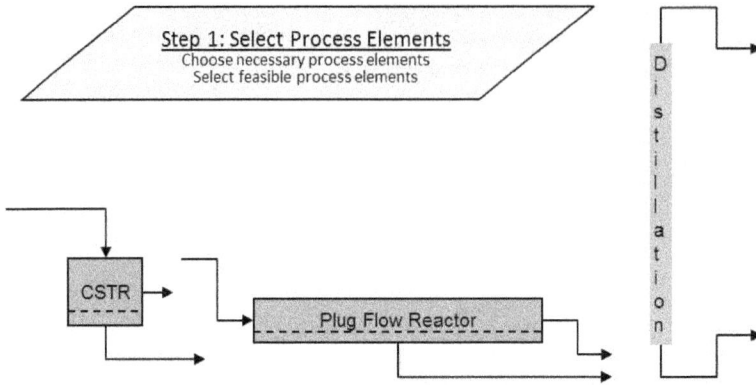

Figure 15: Selection of Superstructure Elements

For simplicity, this example does not consider a recycle of unreacted feed A and B. In a first step, continuous stirred tank reactor (CSTR), plug flow reactors (PFR) and distillation are the three elements for reaction and separation. These three technologies are available as technology elements to create a superstructure as shown in Figure 15.

In step 2, these technology selections are used to configure a superstructure for reaction and separation. Figure 16 shows a simple superstructure for the reaction section. This reactor superstructure possesses alternative process features such as back-mixing/no back-mixing, different ways of feed addition and product removal. The reactor superstructure has to include all relevant, alternate reactor structures that are feasible.

In a next step, the designer composes a superstructure for the separation task.

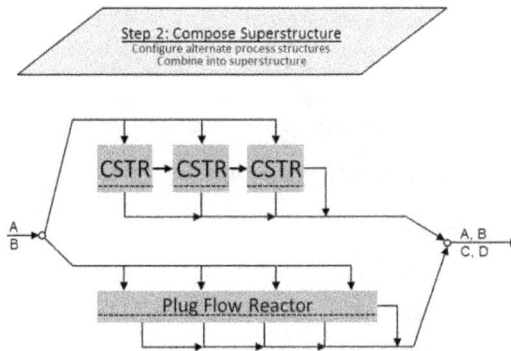

Figure 16: Reactor Superstructure

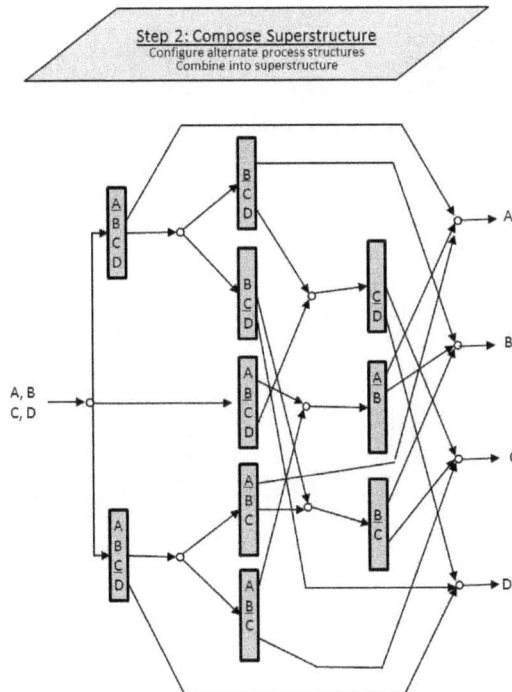

Figure 17: Separator Superstructure

The effluent of the reactor contains the components A, B, C, and D. A separation superstructure has to include the separation sequence and unit operations.

Since distillation is the only technology provided, the superstructure focuses on the separation sequences. The superstructure for a potential separation sequence of a 4-component mixture into the four pure components is illustrated in Figure 18.

Figure 18 combines the superstructure of the reaction and separation sections into a final, complete superstructure, whereas the designer hopes that the superstructure includes the optimal structure for the design task.

There are already 10 alternate structures included in the separation superstructure and 4 alternate structures in the reaction superstructure.

An option for an isotherm or adiabatic operation adds a further dimension to the superstructure. Similarly, the recycle of material from the reactor outlet or separation units back to the different reactors will further expand a possible superstructure.

Assuming there are at least six more separation technologies (such as extraction, membranes, crystallization, adsorption, absorption and filtration) available, the number of alternative structures increases dramatically.

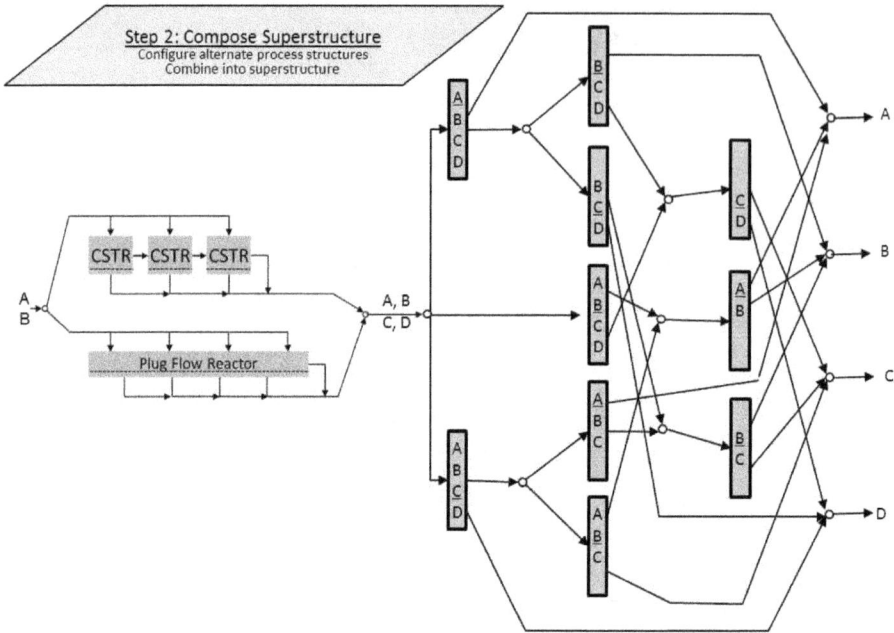

Figure 18: Complete Superstructure

This complexity issue explains why unnecessary superstructure elements that are not beneficial should not become part of the superstructure of course without endangering the purpose of a superstructure to include the optimal structure.

The final step of a superstructure approach for process design covers the superstructure evaluation. The goal is to remove all suboptimal elements until the optimal structure emerges from the initial superstructure.

Figure 19 highlights the optimal structure (dotted lines) embedded in a complete superstructure.

For this simple example, selectivity is irrelevant, only productivity counts. Even an inexpensive, continuous stirred reactor will provide the selectivity performance required. A plug flow reactor or a cascade of several stirred tanks does not result in better selectivity.

A plug flow reactor and a continuous stirred reactor cascade lead to a lower concentration of the feed material A and B compared to a single CSTR. This reduces the later separation effort. The reactor of choice is likely a plug flow reactor or a cascade of CSTRs.

Feed addition and product removal are removed from the superstructure. The single CSTR solution is excluded from the superstructure as well. A simple plug flow reactor remains as reactor choice, although a CSTR-cascade is possible as well.

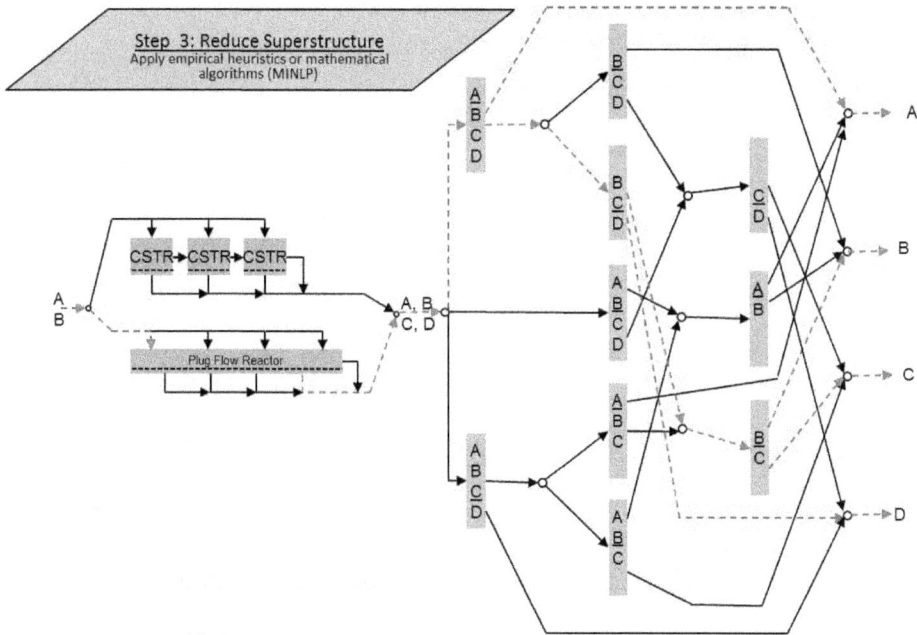

Figure 19: Superstructure with embedded, optimal Structure

A decision on the separation sequence appears difficult based on the limited information given (only boiling points). Probably the difficult separation B/C is performed last (with only these both components involved).The sequence does not involve separation A/B or C/D that are removed from the superstructure. A decision whether component A or D is recovered first is not possible due to sufficient information. A possible sequence could start separating A, then D followed by the most difficult separation last.

Figure 20 shows a process structure consisting of a plug flow reactor and tree distillation column system. The separation sequence performs a split between A and B, C, D first. Next D is recovered as bottom product. Finally, B is separated from C.

Figure 18 describes the superstructure as the result of step 2. Figure 19 shows the optimal solution embedded into this superstructure. Figure 20 summarizes the optimal structure after completing step 3 of a superstructure approach.

How to find an optimal structure in a superstructure is discussed in detail later. This example only highlights the features of a superstructure methodology.

A superstructure approach is only successful, if the superstructure developed in step 2 contains the optimal structure and this optimal structure is found during step 3. To a large extend, the superstructure approach transfers the "critical" design step from structure configuration to structure analysis.

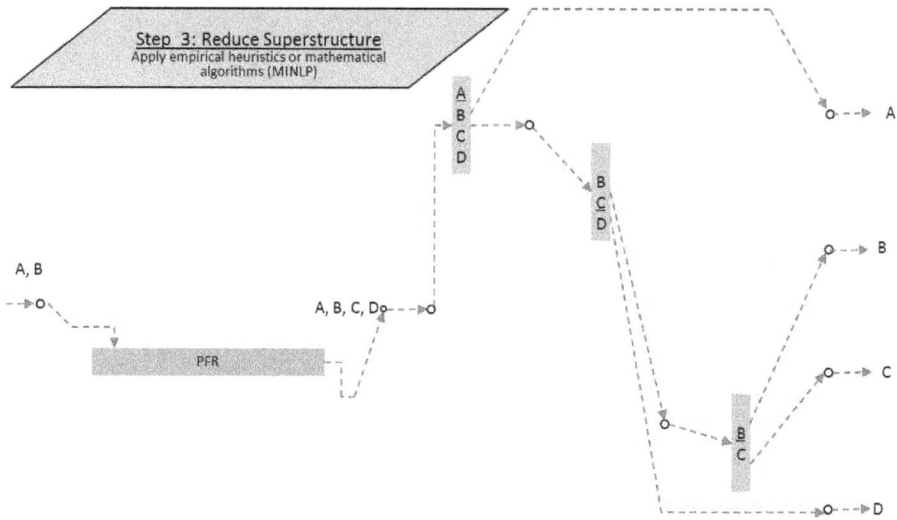

Figure 20: Final Process Structure derived from Superstructure

This superstructure represents the starting point containing all potential solutions. During process design, redundant, inferior operations will be removed from the superstructure leading to the final design.

The first challenge tor this approach is not to overlook the structure that finally turns out to be superior when designing the superstructure. The second challenge to the superstructure approach is to find the optimal solution embedded in a complex superstructure. The selection of an optimal structure is particularly difficult, if only limited data on the process are available.

What tools are available to perform these selections successfully? These selection tools are equally important for a superstructure or a hierarchical design approach.

Design Tools – Algorithms or Heuristics?

The hierarchical approach follows a bottom-up, sequential concept, while the superstructure approach is a top-down, simultaneously optimizing concept. The superstructure focusses on a solution space including all feasible solutions. The sequential, hierarchical approach starts with the core operation – normally reaction – adding new process elements gradually.

Generally, the superstructure approach including the optimal structure within all alternatives becomes too complex to be handled successfully. The hierarchical approach appears to be straight-forward and technically manageable – but may not lead to the optimum, since an early discarded structure element together with a later added structures might represent the optimal solution. A hierarchical approach, however, will miss this solution due to a bad decision earlier.

Both mental concepts – superstructure or hierarchical approach – require tools to make design decisions whether to remove an inferior structure element or add a structural element in a next step. The two tools available for these tasks are

- Mathematical Algorithms
- Empirical Heuristics

Mathematical algorithms and empirical heuristics represent the main tools for developing the solution regardless whether the starting point is a superstructure or the initial level of a hierarchical approach. These tools can be applied manually and with computers as well.

Mathematical Algorithms

Process design using strict mathematical algorithms follows the well-known concept of model generation, process simulation, and process optimization. The optimization goal is the process structure.

The optimization function is very often a cost objective:

Minimum Cost = f (structural and process parameters)

The simulation model requires

Process models (mass/energy balances, process/equipment models ...)

Data (kinetics, materials, cost, financials, etc.)

Mathematical Optimization Algorithms

An optimization algorithm is necessary to find the optimal process structure.

Process modeling, simulation and optimization for a <u>given</u> process structure is a standard task extensively taught in chemical engineering courses. Various non-

Integer Variable Y_i:

$$Y_1+Y_2+Y_3+Y_4+Y_5=1$$

$Y_i= 0$ oder 1 $i = 1, 2, 3, 4, 5$

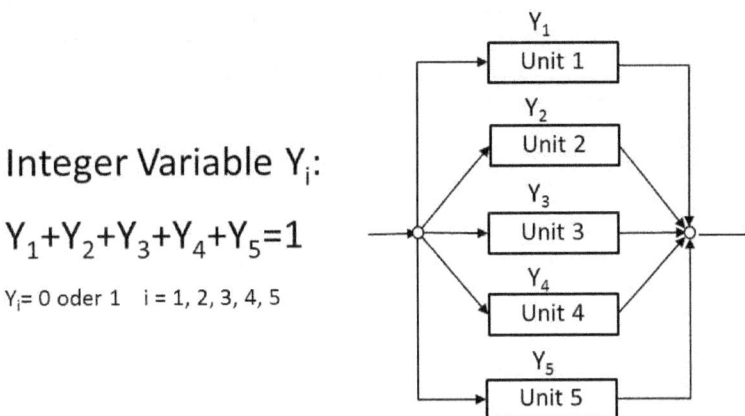

Figure 21: Mixed Integer Non-linear Programming (MINLP)

linear programming algorithms efficiently do parameter optimization for flow rates, stream composition, equipment sizes or process temperatures and pressures.

Process synthesis to derive a novel process structure from a superstructure adds another dimension to mathematical programming. Structural optimization requires *discreet structural parameters*.

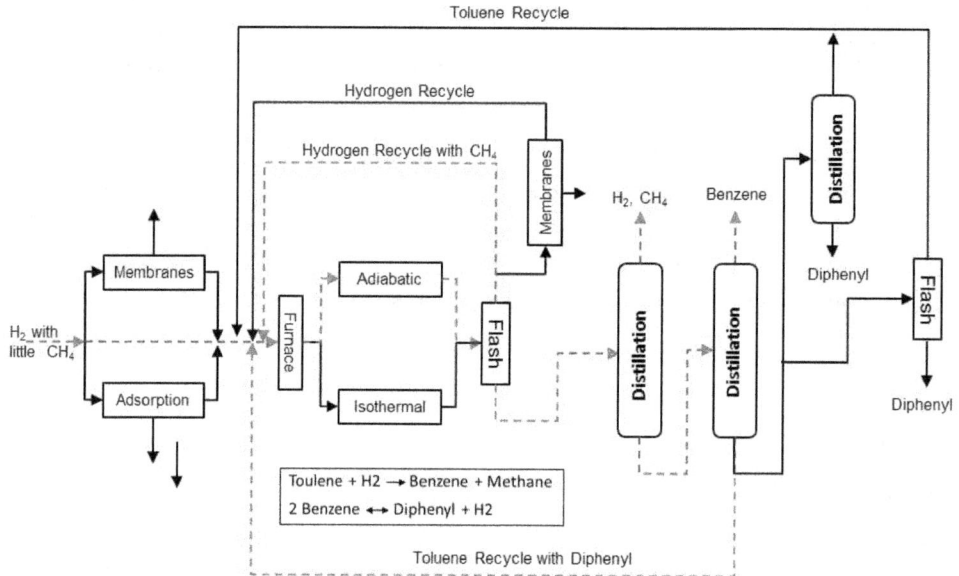

Figure 22: Process design with mixed integer non-linear Programming

In addition to the continuous parameters, integer parameters are assigned to structural elements of the process. These integer parameters can only have two values 0 or 1. If a structural parameter is 1, then this structural element remains in the process model. If the integer becomes 0, the algorithm removes the structural element from the model (Figure 21).

During optimization, a mathematical algorithm needs to discard structural elements of the process model to proceed to an optimal solution. This adds "discontinuities" to the optimization progress.

Non-linear programming algorithms (NLP) are expanded to deal with integer structural parameters.[4] The mathematical tool is called mixed integer non-linear programming (MINLP).

In Figure 21 an integer variable Y_i is linked to five alternate process elements (for example different reactors). Adding the boundary condition $\sum Y_i = 1$ forces the

---

[4] Biegler, L. T. (2014). CIT, pp. 943-952

system to select only one process element during structural optimization. Operation cost is minimized depending on continuous process parameters and discrete structural parameters.

Applying MINLP to a superstructure shown in Figure 19 becomes a very complex task. In addition to the mathematical difficulties, this programming approach requires detailed, upfront information about all structural elements.

The superstructure approach using mixed integer non-linear programming to select an optimal process structure has remained in the ivory tower of academia up to now and not reached industrial practice, yet.

Nevertheless, thinking in superstructures can add insight to process design. Smaller synthesis problems are likely to be successfully solvable with a MINLP-approach.

Process synthesis with a MINLP-approach is applied to manufacturing benzene from toluene and hydrogen converting the superstructure into the final process structure (dotted lines) shown in Figure 22. In this example, an isothermal reactor was removed from the superstructure in favor of an adiabatic reactor design. Since methane is generated in the reactor as a by-product removed from the recycle stream in a separation unit, an additional separation unit to remove the "impurity" methane from the hydrogen feed is not necessary. Diphenyl is recycled to extension due to the reaction system including an equilibrium reaction. Redundant separation units and recycles are removed from the superstructure as well. [5]

Empirical Heuristics

Empirical heuristics offer an alternative approach to mathematical modeling and simulation to derive an optimal structure from a process superstructure. Heuristic rules describe empirical, qualitative knowledge about process synthesis. Heuristic rules cover methodology aspects, design rules, and process facts.

Heuristics possess the following structure:

If "fact=true", then perform "action"

A heuristic rule links a design fact with an action. Heuristics cover various areas:

Method Rules:

If "various rules are applicable", then "start with prohibitive rules"

Design Rules:

Reactor Selection Rule:

If "back-mixing lead to by-products", then "use plug flow reactors"

Separator Selection Rule:

---

[5] Figure 22 is a simplified version to demonstrate the methodologies. More details can be found in the literature (Douglas, 1988), (Grossmann. (1989). Comp.Chem.Eng., pp. 797-819)).

If "mixture is liquid at ambient pressure", then "try distillation"

Distillation Sequence Rule:

If "component is sensitive to temperature", then "separate component first"

Heuristics to tackle a process design task represent a comparably simple, rule based thinking. Research in artificial intelligence has made significant progress with respect to knowledge management, reasoning, and learning. Nevertheless, process design still applies heuristics in a straightforward, sequential approach.

Heuristics are applicable in a superstructure or hierarchical design approach to find the optimal process structure. Process design in reality uses mathematical algorithms and empirical heuristics simultaneously.[6]

A simple example illustrates the application of heuristics to derive the optimal process from a superstructure. Figure 23 shows the process superstructure. Limited process information and six heuristic design rules are available to make the design decisions:

Process information:

A + B → C   : C desired product

C + B → D   : D undesired by-product

Reactions   : slow

A, B, C, D   : liquids at ambient pressure

A   : temperature-sensitive

B/C   : difficult separation

Heuristic design rules:

If "reaction is slow", then "try stirred tank reactor" (1)

If "back-mixing reduces yield", then "do not use stirred tank reactor" (2)

If "components are liquid at ambient air pressure", then "use distillation" (3)

If "component is temperature sensitive", then "remove first" (4)

If "rules are contra dictionary", then "select prohibitive rule first" (5)

If "separation is difficult", then "perform last" (6)

The structure evaluation using heuristics starts with the reaction section. The reaction system indicates that the reaction is slow and back mixing of product C with raw material B supports the formation of by-product D detrimental to process yield.

---

[6] *Actually, the term "rule-based and mathematical algorithms" describes both design concepts more precise than "mathematical algorithms and empirical heuristics – both are algorithms.*

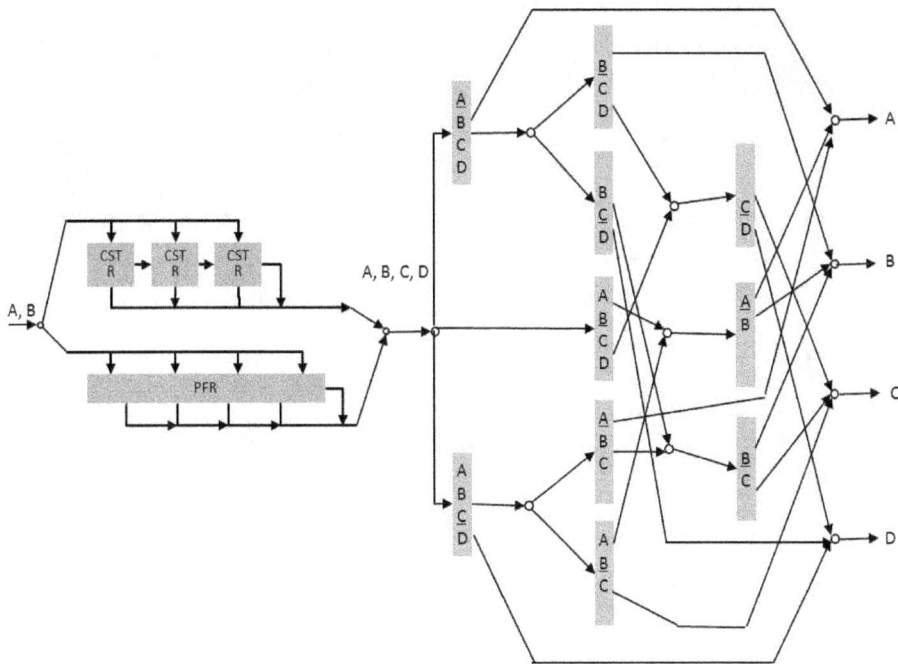

Figure 23: Initial Superstructure for heuristic Approach

Heuristic 1 recommends a stirred reactor for slow reactions, while heuristic 2 prohibits stirred tank reactors, if back mixing leads to more by-products. Heuristic 1 and 2 request different designs.

Based on heuristic 5 that describe how to apply rules, the prohibitive rule 2 is applied with a higher priority than rule 1 leading to the selection of a plug flow reactor.

Since the components A, B, C and D are liquids at ambient air pressure. Rule 3 recommends distillation as separation technology. Heuristic rule 4 prefers the early removal of A, since it is sensitive to temperature. The immediate separation of A from B, C, and D avoids multiple processing of A at higher distillation temperature necessary in other process design options.

Separation of B from C is difficult. Rule 6 suggests that the separation of B and C is best done last. Other components cannot interfere and further complicate the difficult separation of B and C.

If A is removed from the mixture first and B and C are separated last, the second separation step requires a separation of D from B and C.

Figure 24 summarizes the application of heuristics to the superstructure removing inferior solution elements to derive an optimal solution (see dotted lines). Figure 25 shows the final process structure extracted from the superstructure.

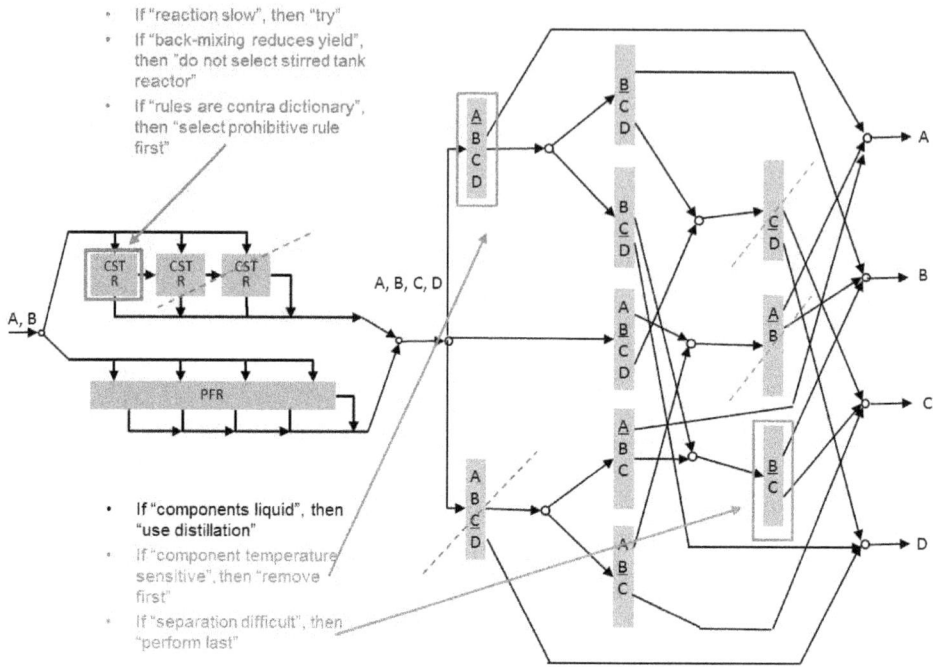

Figure 24: Superstructure evaluated with Heuristics

Of course, a realistic example requires a more sophistic approach. Nevertheless, this example illustrates the benefit of heuristics to process design.

Mathematical and heuristic algorithms are often applied to a design problem simultaneously. Many hybrid concepts combining rigid modeling with heuristics have been proposed in the past. Both approaches – mathematical and heuristic algorithms – can be implemented on computers.

The ultimate design tool for a process designer represents an automatic, computer-based methodology. This automated design should convert the information on the process task directly into an initial process structure. To a large extend this effort failed in the past due to limited computer power for huge problems – particularly if you take into account that such an automated design system has to cover dynamic constraints during the lifetime of a new process.

Progress in artificial intelligence combined with increased computational power will certainly enable automated process design systems. Nevertheless some concern remains, whether a straightforward methodology combining artificial intelligence and computer power will solve the challenge for large complex, emergent systems in an uncertain environment.

In early design stages, too complex models conceal simple facts. Extensive simulation regularly generates a false sense of security. This is particularly true, if input data for the complex models are limited or not validated.

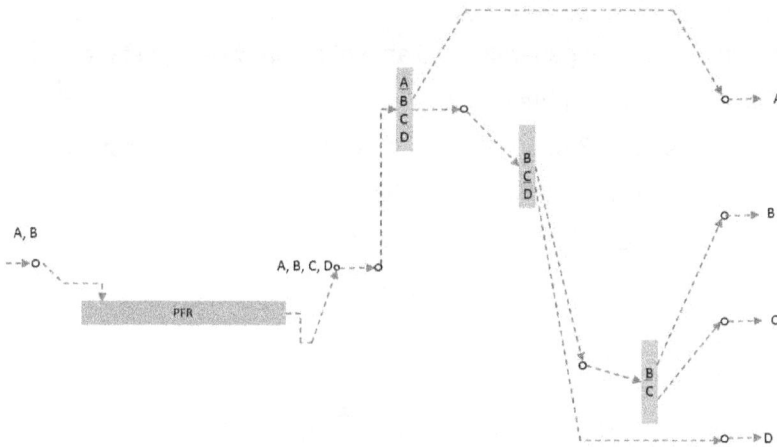

Figure 25: Final Process Structure

Intelligently simplified models focusing on important design features are indispensable for complex topics. Only "simple" models create true insights - particularly valid to handle complex issues in early design phases.

Some general conclusions follow with respect to process synthesis:

- In a mature business like the chemical industry, a process designer does generally not develop processes from scratch.
- Broad knowledge on processes and technologies exists in empirical heuristics. This facilitates and favors a rule-based approach.
- Process designers follow an approach actually combining thinking in superstructures and hierarchies applying empirical knowledge organized in heuristic rules.
- The basic decisions to select a process structure are generally made using a very limited list of heuristics.
- Process designers tend to decide on a process structure very fast and subsequently spend a lot of time on computational simulation and experimental studies on this process structure (methodological biases).
- Since empirical knowledge is gathered in humans, even or particularly experienced design engineers are in danger of focusing on their preferred heuristics (cognitive biases).
- This behavior regularly leads to inferior design structures with extensively optimized operational process parameters – still missing the truly best solution.

A design methodology has to take into account this behavior. Chapter 3.2.2 introduces the concept of targets and benchmarks to counteract theses biases. Chapter 3.6 discusses economic and financial concepts to ensure profitability of a process design. Technology excellence is necessary, but economic profitability finally decides on a process design.

Design Case 1 - Methodologies[7]

Only limited information is generally available at an early design stage:

Process Goal: Recover a pure product

Reaction: Feed → Product   catalytic, slow, exothermic Reaction

Constraints: Feed contains little impurities

Boiling Point: Product>Impurities>Feed

Components are liquid at ambient conditions

Feed material expensive

Product temperature sensitive

Impurities react with product at high temperature

In addition to the process information, some heuristics are provided to guide the design process:

If "reaction is slow", then "try stirred tank reactor cascade" (1)

If "back-mixing reduces yield", then "do not use stirred tank reactor" (2)

If "component is liquid", then "use distillation" (3)

If "component is temperature sensitive", then "remove first" (4)

If "rules are contra dictionary", then "select prohibitive rule" (5)

If "separation is difficult", then "perform last" (6)

If "impurities can react with product", then "remove early" (8)

If "feed material is expensive", then "recycle to convert completely" (9)

Process design is illustrated for this design task using a hierarchical approach first.

In step 1, based on the reaction information – straight forward, no by-products, slow – a stirred tank appears to be a good reactor choice. Since it is a slow reaction, a reactor cascade is probably necessary to provide sufficient reaction time (Figure 26).

In step 2, recycle of feed is proposed to convert unreacted feed into product. Rule 9 demands feed recycle, since expensive feed must not be wasted (Figure 27).

In step 3 of the hierarchical design methodology, first a sequence for the separations is selected. Rule 8 requires an early removal of impurities. The earliest option to remove the impurities from the system is to remove the impurities already from the feed before a catalytic reaction takes place. This decision reduces the

---

[7] *Design cases are used in this book to demonstrate design methodology feature in selected, adapted examples through the various book chapters.*

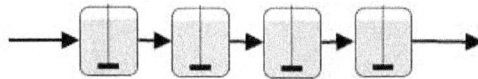

Figure 26: Reactor Design – Step 1

separation of the reactor effluent mixture to a simple separation of feed and product.

Since all components are liquid at ambient conditions, rule 3 recommends distillation as separation operation technology.

Figure 28 illustrates the process structure derived by a hierarchical approach using heuristic rules.

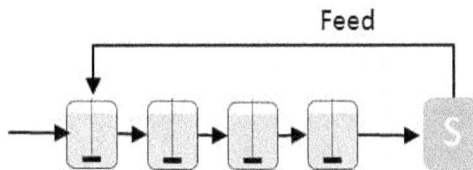

Figure 27: Recycle Decision – Step 2

The hierarchical approach appears to lead to an acceptable solution. However, is this process structure the best?

The input data can also be used to apply the superstructure approach. The basic unit operation elements include continuous stirred tank and plug flow reactor and separation device such as distillation, extraction, or chromatography. The

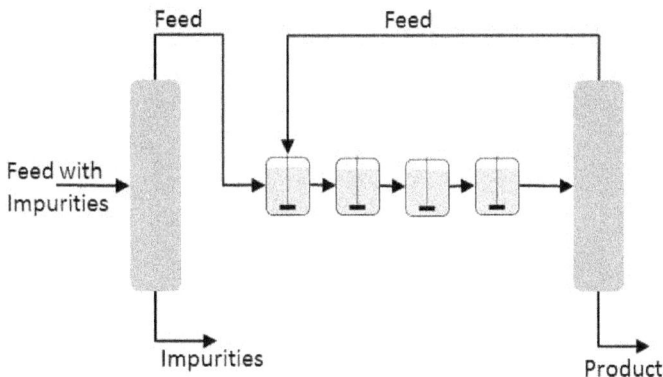

Figure 28: Separation Sequence – Step 3

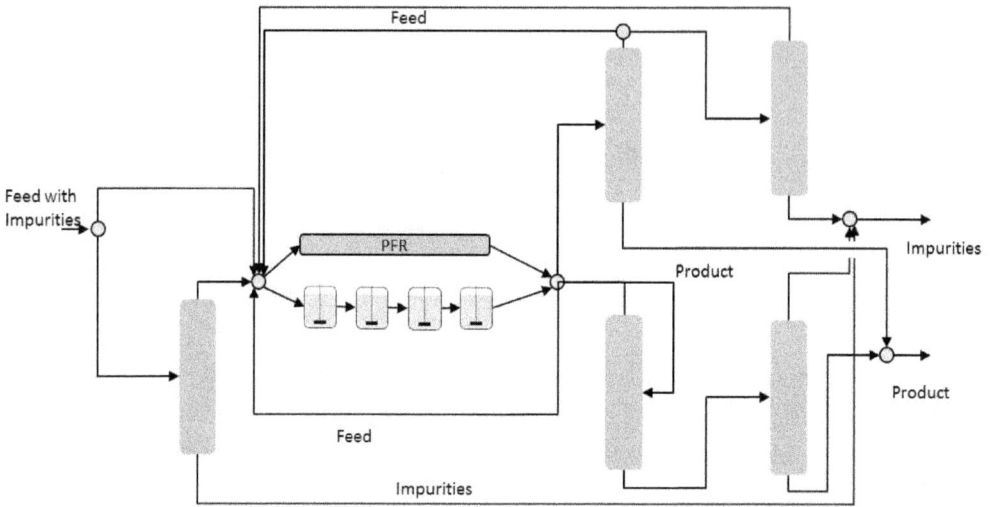

Figure 29: Superstructure with embedded, optimal Solution

superstructure for this design task contains 11 elements linked by material flows. Figure 29 describes the resulting superstructure.

The same logic that was applied to guide the hierarchical approach can be used to remove inferior elements to derive the optimal solution. Rule 8 – remove impurities early – recommends at the first junction in Figure 29 to select the separation unit to remove impurities. At the second junction, rules 1 leads to a stirred tank reactor cascade. The reactor effluent contains only feed and product. Rule 9 demands a

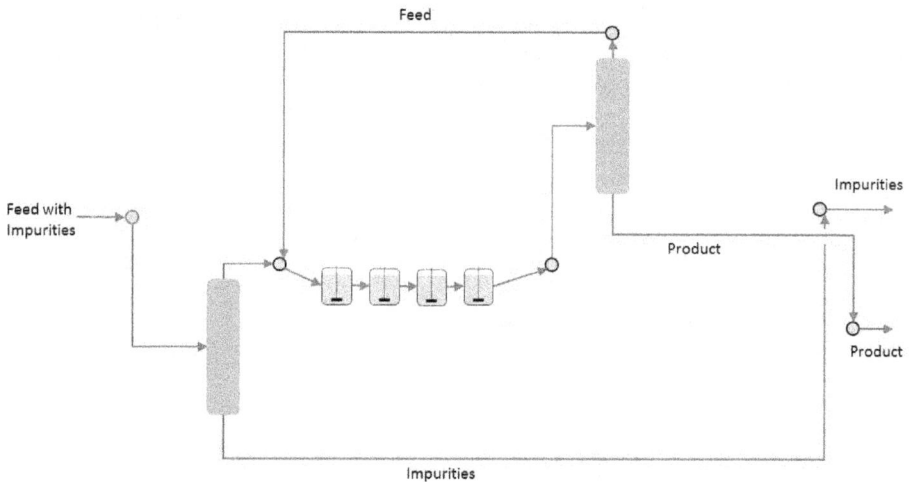

Figure 30: Process derived from Superstructure

feed recycle to avoid feed loss. A separation of feed from product by distillation is the result of rule 3. Figure 30 highlights this process structure.

This simple case study already indicates that the development of an initial process structure follows an approach actually combining thinking in superstructures and hierarchies with application of empirical knowledge organized in heuristic rules.

Design Software – An indispensable Tool

Process design heavily relies on software to model, simulate, and optimize design alternatives. These process simulators are part of an integrated computer-aided design package that contains computer-aided engineering software, computer-aided design systems, and process simulation software.

Software manages flow-sheet information, mass/energy balances, property data, equipment specification, and cost data are jointly during all the design phases from process development to plant operation.

Process simulators consist of various building blocks:

- Unit Operation Library
  - o   Models for reactors (stirred tank, plug flow, fix bed, membrane reactor etc.)
  - o   Models for separators (distillation column, Extractors, Chromatography etc.)
  - o   Models for auxiliary equipment (pumps, splitter, pipes etc.)
- Reaction Kinetic Library
  - o   Models for different reactions types
- Material Property Library
  - o   Material data (density, boiling/melting points etc.)
  - o   Mixture data (solubility, vapor-liquid equilibrium etc.)
- Cost Library
  - o   Equipment cost, raw material cost, utility cost
- Mathematical Algorithms for mass balances, simulation and optimization

Process simulation generally follows a certain procedure:

- Use the most promising process structure to build a simulation model
- Select unit operations and equipment from software library
- Define streams and recycles between equipment
- Specify input data (feed rates, equipment size, tray number, heat exchanger...)
- Define initial process parameters (temperatures, split ratios ...)
- Validate simulation model
- Calculate conversion, selectivity and yield
- Establish mass and energy balances
- Calculate manufacturing cost and capital investment
- Evaluate different process and equipment parameters
- Use to compare alternate process structure and optimize parameters

A comprehensive overview on process simulation and simulator are given in the literature (Seider, Seader, & Lewin, 2004).

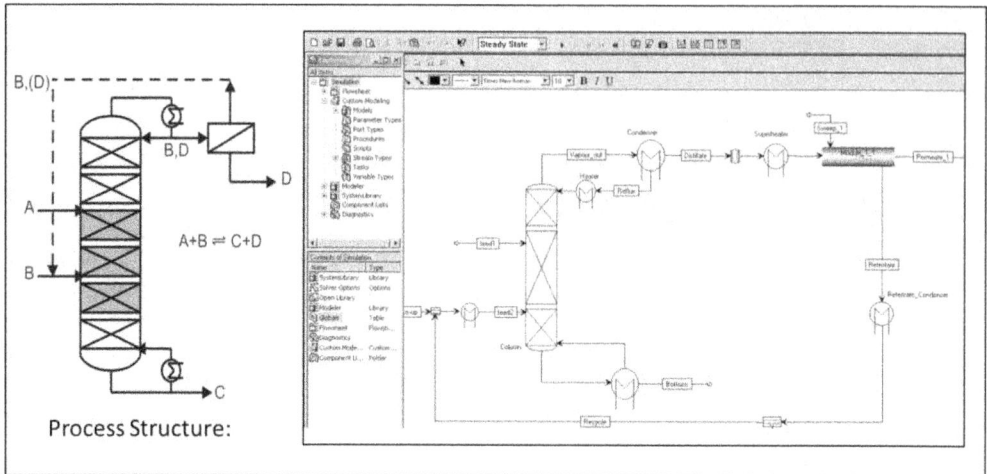

Figure 31: Process Simulation Software (Source: TU Dortmund)

There are various process simulators commercially available such as Aspen Technology, HYSYS/Unisim Design, gPROMS, CHEMCAD, PRO/II.

While these software packages are full service systems with all relevant components, some systems are suited for a specific task. Heat integration to minimize energy consumption is accomplished using dedicated software as SPRINT, HEAT-int, Speed-up, and SuperTarget.

Furthermore, there is a wide range of general-purpose optimization software available. GAMS is a modeler for linear, non-linear, and mixed-integer problems. LINDO was one of the first commercial tools to handle mixed-integer optimization.

There is a huge variety of software tools commercially available. Figure 31 describes the AspenTech – Simulator.[8] Klemes gives an overview on software available to process design (Klemes, 2011).

Many major chemical companies use in-house process simulators in addition to commercial simulation software, although the in-house simulators are gradually replaced by commercial software. The process data (equipment, materials, parameters etc.) are the important intellectual property of chemical companies. The software is of minor relevance. Software development is not the core business of chemical companies.

---

[8] www.aspentech.com

2.3 Design Methodology – A holistic Approach

Process design requires a holistic approach. This chapter translates the earlier findings into a comprehensive design guideline, while chapter 3 discusses the methodology details in detail.

The process design methodology is divided into three main modules: Synthesis/Analysis, Intensification/Integration, and Modularization/Standardization. The modules Targeting, Economics, and Future Scenarios provide tools to assess different design solutions. Figure 32 schematically outlines the design procedure.

Starting point is the definition of the process task in terms of input data such as product volumes, synthesis pathways, raw materials, and available process technologies.

Process synthesis generates process structures, while process parameters are specified in detail during *process analysis.* Process synthesis and analysis interact to create an initial process structure with a first set of process parameters. This initial process is then optimized using *targeting and economic evaluations. Process intensification* and *process integration* form special areas in process design.

Intensification primarily focuses on unit operations evaluating whether intensification of heat and mass transfer can improve process efficiency. *Process integration* studies process structures for system synergies. While in the early days, process synthesis and analysis dominated the process design arena, today real progress also demands a focus on intensification and integration benefits – particularly for the mature chemical industries.

Finally, *Modularization and Standardization* select an appropriate capacity expansion and cost reduction strategy. Modularization is a valuable tool to enable

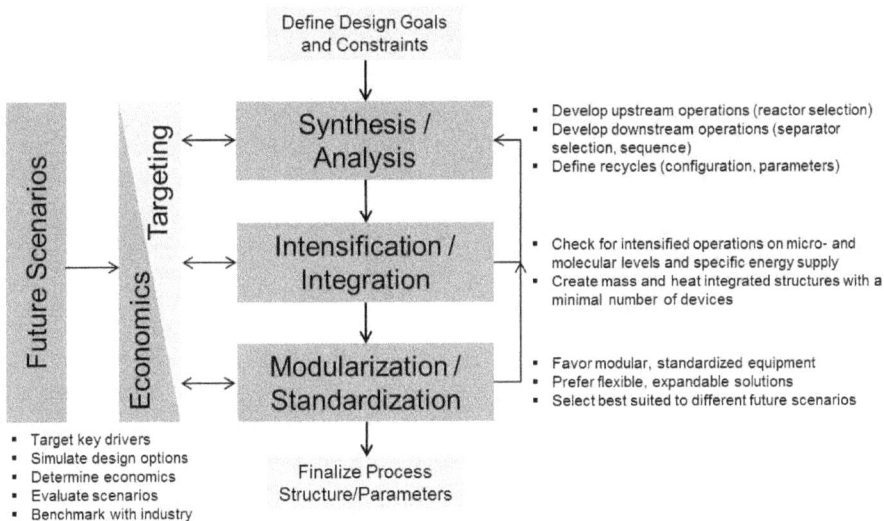

Figure 32: Process Design Methodology

a stepwise capacity expansion to meet product demand flexibly. Process design has to prefer standardized solutions to avoid excess investment cost of tailor-made solutions.

In general, the methodology outlined in Figure 32 will be applied several times in a trial and error approach to find the optimal structure and parameters for the process.

The methodology is briefly demonstrated to explain the methodology logic and features.

Design Case 2 – Demonstration Example

Let us assume that a fine chemical is to be produced in a two-step reaction as described in Figure 33.

The first reaction is fast and exothermic requiring a significant dilution with a

Process Data:

$$A + B \rightarrow C \qquad \text{fast, exothermic}$$
$$C \rightarrow D \qquad \text{undesired by-product}$$
$$E + C \leftrightarrow F + G \qquad \text{catalytic equilibrium reaction}$$

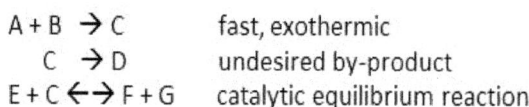

Raw materials: A, B, E
Product: F
By-Products: D, G
Boiling Point: $T_D < T_A < T_B < T_C < T_F < T_E < T_G$

Figure 33: Demonstration Case - Reaction Scheme

solvent or excess of raw material to avoid formation of a low boiling by-product D due to local hot spots in the reactor. A is added in excess to the first reactor.

Component E together with a suspended catalyst is fed to a second reactor. Since the reaction 2 leads to the equilibrium, the resulting mixture needs to be separated and raw materials C and E are recycled to reactor 2.

The resulting initial flow sheet of the *Process Synthesis* step is shown in Figure 34 consisting of two reactors and two separation systems. The second separation system is a separation sequence with a catalyst centrifuge and two distillation columns.

During *Process Analysis*, the process flow sheet is optimized with respect to process parameters searching for the minimal cost of goods and/or capital investment.

Reactor 1 is used to demonstrate the logic of *Process Intensification*. Can intensified mixing and heat removal improve the reaction 1? The production of by-product D is caused by local hot spots in the reactor due to insufficient heat removal. Micro-reactors with excess heat exchange area are a promising solution

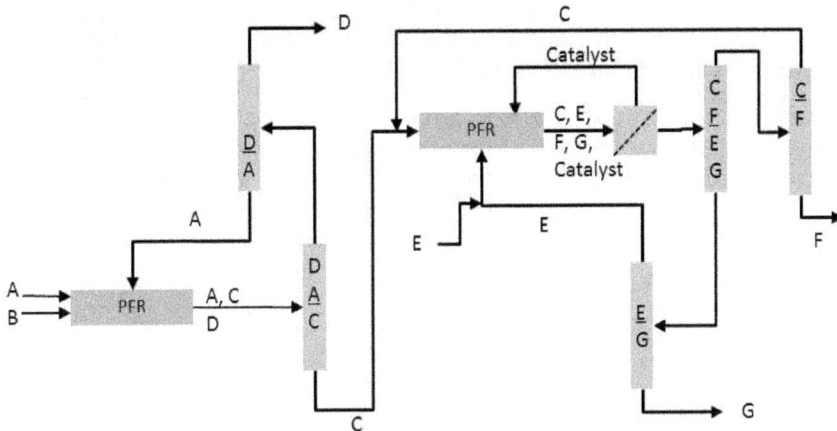

Figure 34: Demonstration Case - Initial Flow Sheet (after Process Synthesis)

to remove heat efficiently. Catalyst recycle can be avoided, if the catalyst is coated to the walls of the micro-reactor.

Equilibrium reactions cry for solutions that influence the equilibrium positively. Preferably, the desired product is removed from the reaction zone, while educts are returned to the reaction zone continuously.

Process Integration realizes this by combining reaction and separation in one reaction column. The insertion of a dividing wall into the column allows product removal without contaminating the product F. A reaction dividing wall column performs equilibrium reactions particularly efficient, provided data allow the integration. Heat integration offers additional optimization potential, when the exothermic reaction heat can be used to heat the column evaporator.

Finally, a decision is required whether the process is to be constructed as a single train, tailor-made plant or a modular approach is chosen to allow a stepwise implementation suited to a slow demand growth. Modularization and standardization are decisive to manage cost and create flexibility (*Process Modularization*).

Obviously, the procedure is not straightforward; the different steps are likely to be repeated several times. Simulation, targeting and benchmarking accompany synthesis, intensification, integration and modularization during process design.

This methodology is outlined in detail in chapter 3 with the following topics:

Process synthesis aims at deriving the basic process structure and parameters for the plant output desired: *Process Synthesis – it is about Structuring Complexity.*

To improve the probability to come up with superior solutions for complex problems the methodology applied requires a target to benchmark the design progress. This target must not be derived using the methodology applied. The targeting process, therefore, derives a target from first principles for different scenarios identifying key drivers for alternate boundary conditions. A comparison of design results with such a target supports process design and decision-making: *Process Analysis – It is about Decision Drivers, Targets, and Scenarios.*

Of course, technology optimization and innovation present the core of any methodology to design chemical processes and plants. A special focus needs to be given to intensification and integration of the processes, since unit operations have already been far developed in the mature, processing industry: *Technology*

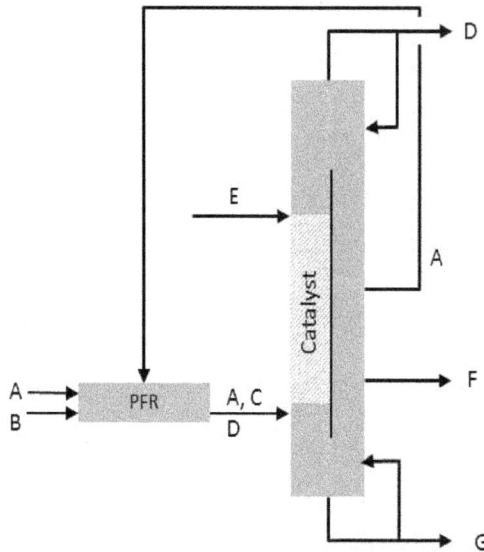

Figure 35: Demonstration Case – Intensified and integrated Process

Opportunities – It is about Process Intensification and *System Opportunities – It is about Process Integration.*

Predicting the long-term impact of a process design on sustainability represents a particular challenge. Under uncertainty to bet on one seemingly optimal solution might not be the best choice. Selection of a design whose performance is acceptable in various scenarios might be the preferable solution. Finally, adaptable or expandable designs are superior to onetime, big scale investment: *Modularization & Standardization – It is about Flexibility, Robustness, and "No-Regret" – Solutions.*

Finally, it is particularly important to avoid biases and emotional traps when straightforward procedures are not applicable. Engineers act in a framework of experience, preferences, and emotions. Minds affect motivation of and interaction in teams enhancing or preventing creative, novel options during early project phases. Behavior is about people: *New Frontiers – it is about Open Minds.*

3   Smart Methodologies – Targets, Technologies and Systems

Chapter 2 gave a general overview on process design methodologies illustrating the procedure, strategies, and general tools. Chapter 3 focuses on methodological aspect in more detail. Further information on process design methodologies can be found in various textbooks (Smith, 2004), (Seider, Seader, & Lewin, 2004), (Douglas, 1988).

3.1   Process Synthesis – It's about Structures

What distinguishes the search for the optimal structures from the search for the optimal parameters?

Provided a process structure is available, this structure can be evaluated with respect to technical and/or economic criteria. For example, the search for the optimal parameters for a given process structure can be pursued by modeling, simulation, and optimization. Alternative structures can be compared, different parameter sets analyzed for the optimal one. Even if the starting parameter set is far away from the optimal solution, this solution is likely to be found when sophisticated mathematical or experimental strategies are applied to the problem.

In contrast to parameter optimization, the designer is never certain that there is no better process structure feasible. Evaluation of many process structures reduces the likelihood that a designer overlooks the best process structure, but is no guarantee – at least if the design problem is complex. If the initially chosen process structure is sub-optimal, comprehensive parameter optimization during basic and detailed engineering will not correct this disadvantage.

There are no fool-proven, straightforward methods for the creative structure search

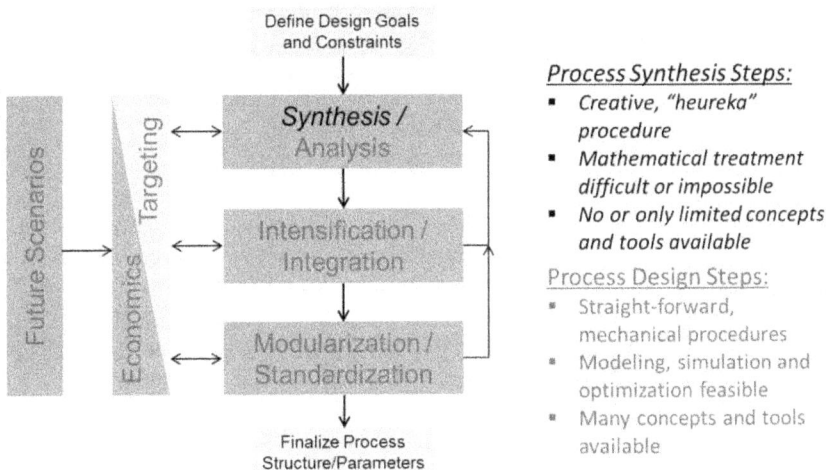

Figure 36: The Uniqueness of Process Synthesis

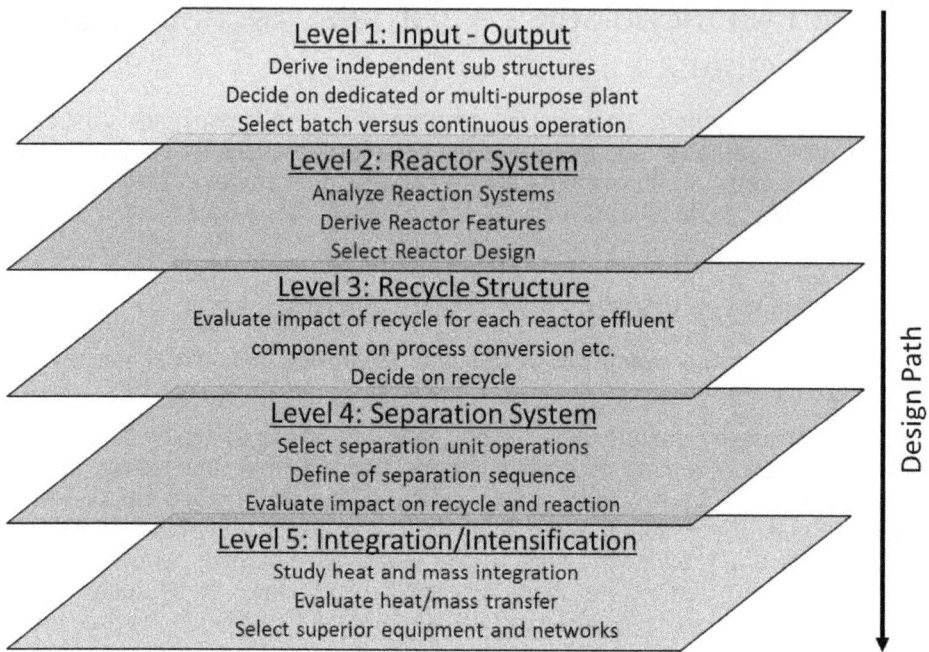

Figure 37: Hierarchical Design Methodology

available.

Two design steps in the design methodology described in Figure 36 represent a truly "creative" challenge for the process designer. Process synthesis and process integration deal with process structure issues.

Process synthesis generates an initial process structure consisting of three major tasks: reactor selection and design, design of recycles / purges, and design of separation / formulation systems. Process integration realizes system synergies by linking unit operations more efficiently.

As outlined earlier a hierarchical design approach enables the designer to handle the structure design task in a simplified way.

Figure 37 illustrates the hierarchical design path. On level 1, the input – output data are defined and the overall process design task is divided into sub problems that are treated individually. In addition, the designer has to decide whether a dedicated or multi-purpose plant and continuous or batch operation is preferable. Level 2 addresses the reactor design. Level 3 decides on recycle streams. Level 4 designs the separation system. Level 5 finally optimizes the system structure.

The next chapters explain the hierarchical design method.

3.1.1   Input – Output Framework

Starting point for a process design procedure is a booklet summarizing the desired product quantities and qualities, the proposed chemistry, available raw materials, utilities and equipment (Figure 38). The prices or cost of these items must be available to allow an economic evaluation of the alternate options.

Of course, the input data should take into account potential developments in the future. Chapter 4.5 addresses design aspects dealing with uncertainty.

Before actually searching for a process structure, the design method selects the operational mode: dedicated versus multipurpose and continuous versus batch operation.

- Product
  - Volume and Price
  - Seasonal demand
  - Special quality requirements
- Feedstock
  - Raw material alternatives and cost
  - Availability
- Chemical conversion pathway, technologies and equipment
  - Selectivity, yields
  - Cost and availability
- Waste
  - Waste Treatment (water, solids, gases)
- Future
  - Novel technologies
  - Future cost/prices of raw material and product
  - Future availability/demand of raw material and product

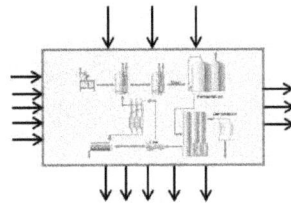

Figure 38: Process Design Input

Input – Output Structure (batch/continuous, dedicated/multi-purpose)

The product portfolio produced in a new plant often defines whether a multi-purpose or several dedicated processes are preferable. From a designer's point of

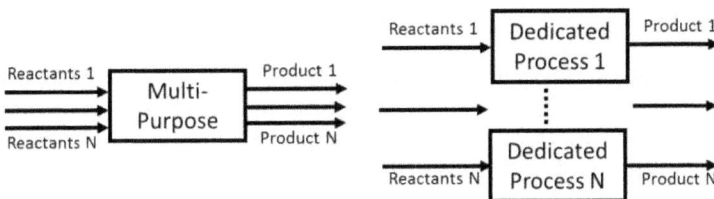

Figure 39: Multi-purpose versus dedicated Processes

view, a dedicated, continuous process always represents the preferred operational mode. Only a dedicated, tailor-made process operated under the optimal conditions provides the highest resource utilization.

If several products need to be manufactured, an optimal process is first designed for each product. Provided the structures and parameters of the dedicated

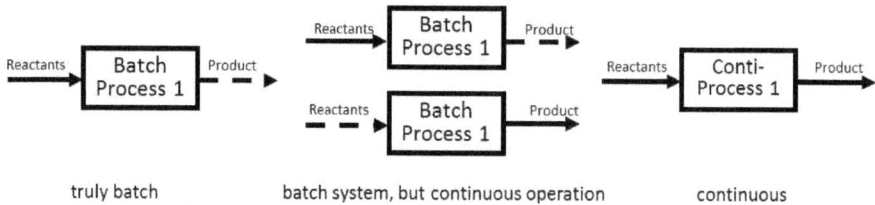

Figure 40: Batch versus continuous Operation

processes possess similar features, a multi-purpose plant can offer advantages (Figure 39).

Similarly, a designer first designs a continuous process with the optimal parameters. This process can always be converted in a semi-batch or batch design, if beneficial (Figure 40). The opposite, however, is not true.

In chapter 5, several heuristics are summarized to make an operational decision. Generally, the nature of the business often drives these decisions.

Subsystem Design

The rationale behind a hierarchical design method is to simplify a complex design task into smaller, manageable design pieces. On level 1 of the methodology outlined in Figure 37, the process designer divides a complex overall system into smaller subsystems that are solved gradually.

The difficulty to select subsystems is to avoid choosing the wrong subsystems. Since these subsystems are optimized independently, an optimal solution due to synergies between subsystems is missed when analyzing only subsystems.

The selection of subsystems is illustrated for a 3-step synthesis:

$$E1 + E2 \rightarrow ZP1 + BP1$$

$$ZP1 + E3 \rightarrow ZP2 + BP2$$

$$ZP2 + E4 \rightarrow P1 + BP3$$

Two feed components E1 and E2 are converted into an intermediate product ZP1 and a by-product BP1. The intermediate ZP1 and feed E3 is then reacted into intermediate ZP2 and a by-product BP2. Finally, E4 and ZPT2 are converted into product P1 and another by-product BP3.

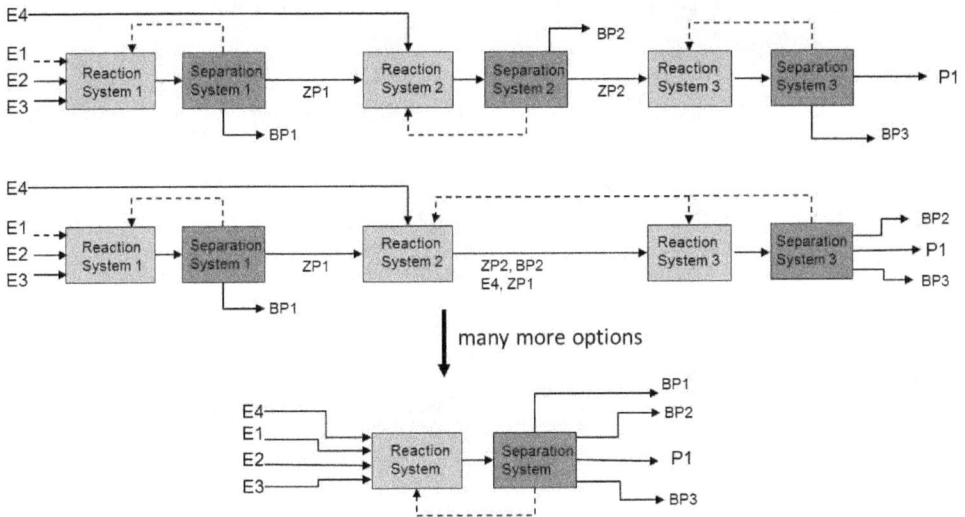

Figure 41: Selection of Subsystems

Figure 41 shows first a standard sequence consisting of three reaction and separation sections. Reaction and separation steps are sequentially performed for each unit. An alternate system could include three reaction sections and only two separation sections. The BP2 is not separated from ZP2. After performing reaction step 3, the by-products BP2 and BP3 are separated from product P1 in a joint separation. Obviously, a direct removal of BP2 can be avoided provided this by-product BP2 does not react with E4 or P1 or make reaction 3 and separation 3 more difficult. Reaction 1, 2 and 3 can be performed simultaneously in one reaction section as shown as third alternative. This is only possible, if reaction conditions (temperature, pressure, and concentrations) are compatible. The same is true for the separation tasks. This process design would only include one reaction and separation section. There are more options to subdivide the design task feasible.

Next, a process structure for each subsystem is derived during evaluation on level 2. Process design could potentially lead to different solutions, if the process designer selects different subsystems in a hierarchical design approach.

On level 1, a decision has to be made which subsystems form the frame for the next design efforts. Heuristics on the selection of subsystems are listed in chapter 5. Heuristics regularly are ambiguous – so use with care!

Design Case 3 – Subsystem Selection

Figure 42 illustrates the application of heuristics to select subsystems. A 3-step reaction is studied:

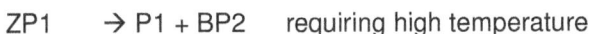

$E1 + E2 \rightarrow ZP1 + BP1$ with best yield at low temperature

$ZP1 \quad \rightarrow P1 + BP2$ requiring high temperature

$$E2 + BP2 \rightarrow BP3 \qquad \text{undesired by-product formation}$$

Rule 16 of chapter 5 (level 1) does not allow a single reaction system if temperature levels to achieve best yield do not fit for both reaction steps. This excludes the third approach with a single reaction section for both reactions in Figure 42.

Rule 18 prohibits a joint separation system 1/2, if components could generate additional by-products. In this case, E2 and BP2 can form another by-product reducing the overall reaction yield.

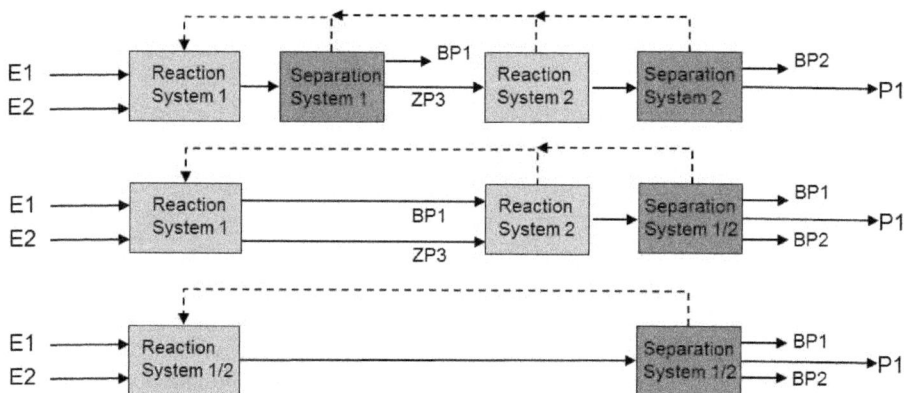

Figure 42: Selection of Subsystems

Nevertheless, the option 2 is acceptable, provided the effluent of reaction system 1 does not include any E2. This could be accomplished, if an excess of E1 is used in the reaction section 1.

Even this simple example shows that every step of a hierarchical design methodology may require trial and error to find the optimal process structure. A hierarchical design approach leads to a design task consisting of three subsystems. On level 2 of the hierarchical approach, process design then starts with an evaluation of the reaction systems 1 and 2.

3.1.2 Reactor Selection and Design

After structuring the design task into subsystems, the next design step is to find a structure for these subsystems.

Conversion, selectivity, and yield characterize the performance of a reaction system with respect to raw material usage. Rate describes the productivity of a reaction system.

Generally, high yield is necessary to minimize raw material usage, while high productivity lowers capital investment. Capital investment (e.g. depreciation) and operating expenses (e.g. raw materials, utilities, labor etc.) determine the cost of goods. Figure 43 summarizes the definition of reaction features.

Reaction Yield:

- Conversion $= \dfrac{Reactant_{used}}{Reactant_{fed}}$

- Selectivity $= \dfrac{Product_{desired}}{Reactant_{used}}$

- Yield $=$ Conversion x Selectivity

Reaction Productivity:

- Rate $= \dfrac{Product_{generated}}{Time \; x \; Volume}$

Figure 43: Reactor Characteristics

Conversion describes the portion of feed fed to the reactor that is converted into product and by-products. The selectivity determines the amount of feed converted into the desired product. Multiplying conversion and selectivity provides the reaction yield. Obviously high conversion and selectivity for the desired product correspond to low raw material usage, while a high rate (productivity) leads to smaller reactors and lower capital investment.

The technical goal is to find the optimal reaction pathway, reaction parameters, reactor type and reactor configuration resulting in a high yield and rate.

Level 2 of a hierarchical design method requires to

- analyze the reactions
- derive reactor features
- select the reactor design.

Reaction Systems

Once more it is more challenging to select the reactor type and reactor structure, then to model the reactor system selected and then to simulate huge sets of input data. The synthesis process is a creative process. After you have selected a structure, you can develop a simulation model and simulate a few scenarios to optimize the process parameters for this process structure.

Reaction analysis has to provide information on the synthesis pathway, kinetic data, and material properties as well. Sometimes quantitative data are available facilitating the synthesis of a reactor section. At an early design phase, a decision on reaction structures often is made with only limited, rather qualitative information.

Fortunately, reactor selection and configuration does not require a complete set of property and kinetic data. In many cases, qualitative information may be sufficient to generate an initial rector design. A comprehensive discussion of reaction analysis and reactor design is given in the literature (Ovenspiel, 1972).

Generally, a reaction system consists of many different reaction steps forming a reaction network. While many reaction combinations in a network are feasible, only a limited set of basic reaction categories are available. Figure 44 gives an overview on these basic reaction schemes.

There are straightforward single reactions systems that convert a feed A and B into a desired product C and a by-product D. The generation of C is directly linked to the generation of by-product D. Of course, there are single reaction systems feasible with more feed components, products, and by-products.

Competing, parallel reactions represent another basic reaction system.

Competing, consecutive systems occur when a product may react with a feed component to create D. Parallel and consecutive reactions compete for the same raw materials generating desired and waste products. In these cases, the selectivity with respect to generation of C and D is affected by the reactor design and parameters.

Consecutive reactions where a feed A is converted into C followed by further reaction to D actually represent a special case of a competing, consecutive reaction.

Equilibrium reactions form an important reaction class. Here the reaction leads to an equilibrium composition of feed A and B, product C, and by-products D that depends on temperature and pressure in the reactor. In this case, feed components will never be consumed completely.

A final reaction category is given with catalytic reactions. Here one or more components are included in the reaction system without being consumed in the reaction. Actually, these components provide a catalyst function and can be other chemicals, enzymes or other microorganisms.

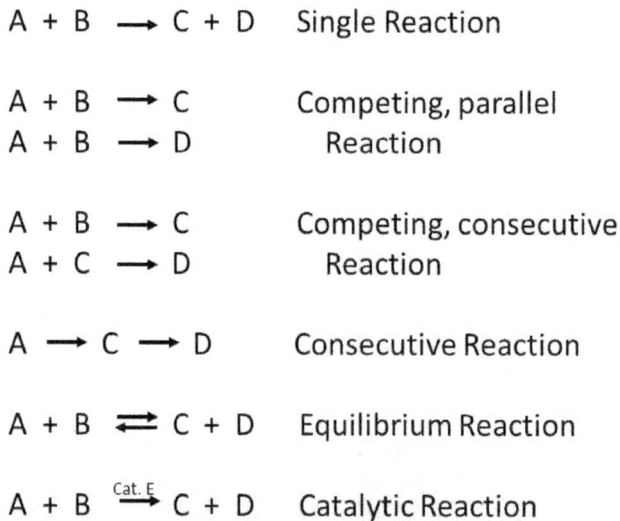

$$A + B \longrightarrow C + D \qquad \text{Single Reaction}$$

$$A + B \longrightarrow C$$
$$A + B \longrightarrow D \qquad \begin{array}{l}\text{Competing, parallel}\\ \text{Reaction}\end{array}$$

$$A + B \longrightarrow C$$
$$A + C \longrightarrow D \qquad \begin{array}{l}\text{Competing, consecutive}\\ \text{Reaction}\end{array}$$

$$A \longrightarrow C \longrightarrow D \qquad \text{Consecutive Reaction}$$

$$A + B \rightleftharpoons C + D \qquad \text{Equilibrium Reaction}$$

$$A + B \xrightarrow{Cat.\ E} C + D \qquad \text{Catalytic Reaction}$$

Figure 44: Reaction Systems

Each of these reaction systems is described by the underlining reaction rate

$$r_i = k_i\, e^{-E/RT}\, c_A{}^a\, c_B{}^b \dots$$

This rate depends on concentrations c_i, rate constant k_i, the activation energy E and temperature T. The impact of the reactant concentration is covered by an exponent a, b ($a=1$ would be a first order reaction and so on). The Arrhenius term describes the impact of the activation energy and temperature on the reaction rate.

Figure 45 summarizes the reaction rate concept for three reaction systems. A system of individual reaction rates characterizes a reaction system.

The first reaction system of Figure 45 directly leads to a product C. Obviously, a maximal rate r_C is desirable to minimize reactor size. Selectivity is irrelevant, since selectivity is not an issue. Higher temperature and concentrations speed up the rate resulting in a shorter residence time to achieve a certain conversion. This means that the reactor becomes smaller and, therefore, less expensive.

The second example shows the rate terms for competing, parallel reactions. Selectivity towards C (desired) is improved when the ratio of r_C / r_D is maximized. This ratio actually depends on the values of the rate constants k, the activation energies E_i, temperatures T and concentrations c_i (and a_i and b_i).

The equilibrium reaction leads to a mixture of A, B, C, D defined by the equilibrium

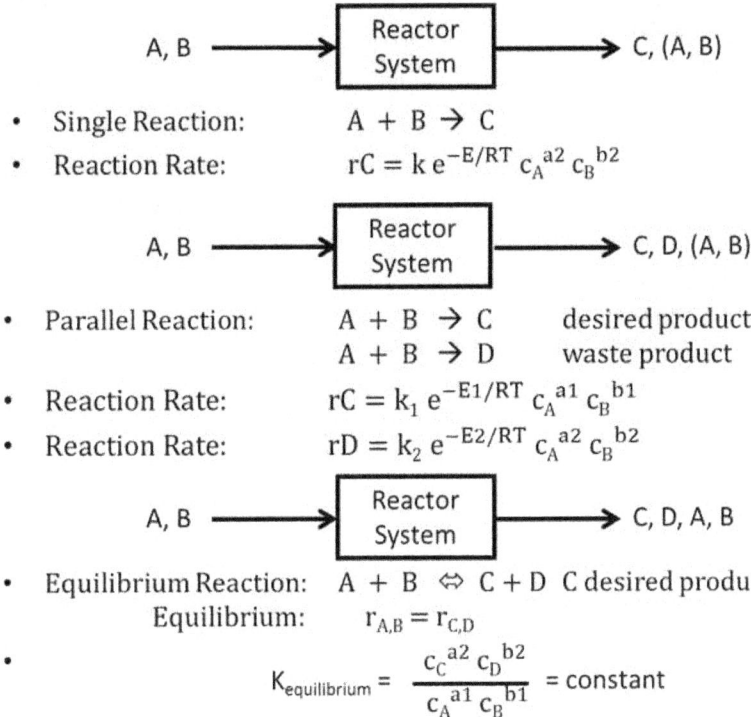

- Single Reaction: $A + B \rightarrow C$
- Reaction Rate: $rC = k\, e^{-E/RT}\, c_A{}^{a2}\, c_B{}^{b2}$

- Parallel Reaction: $A + B \rightarrow C$ desired product
    $A + B \rightarrow D$ waste product
- Reaction Rate: $rC = k_1\, e^{-E1/RT}\, c_A{}^{a1}\, c_B{}^{b1}$
- Reaction Rate: $rD = k_2\, e^{-E2/RT}\, c_A{}^{a2}\, c_B{}^{b2}$

- Equilibrium Reaction: $A + B \Leftrightarrow C + D$ C desired produ
    Equilibrium: $r_{A,B} = r_{C,D}$

- $$K_{equilibrium} = \frac{c_C{}^{a2}\, c_D{}^{b2}}{c_A{}^{a1}\, c_B{}^{b1}} = constant$$

Figure 45: Kinetics of different Reaction Types

constant. For equilibrium reactions, an increased residence time does not change the reactor effluent composition provided the equilibrium state was already reached.

Process synthesis concerning the reactor system deals with selecting the proper concentration profiles and temperature levels that lead to an optimal rate (e.g. productivity) and selectivity (e.g. yield). Pressure is only relevant for gaseous system.

The reaction order (e.g. a_i and b_i) and the activation energy E_i differ for the individual reaction steps. As a result, individual rates are differently influenced by changing temperature levels and concentration profiles.

The following data describe a reaction system:

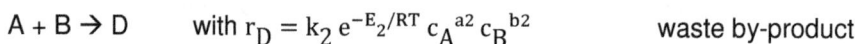

$A + B \rightarrow C$ with $r_C = k_1 \, e^{-E_1/RT} \, c_A{}^{a1} \, c_B{}^{b1}$ desired product C

$A + B \rightarrow D$ with $r_D = k_2 \, e^{-E_2/RT} \, c_A{}^{a2} \, c_B{}^{b2}$ waste by-product

and $k_1 > k_2$, $E_1 = E_2$, $a_1 < a_2$, $b_1 < b_2$

Since the activation energy E does not differ for both reactions, selectivity is not affected by temperature. Productivity improves with higher temperature.

Increasing concentrations of A and B enhances r_D more than r_C. Selectivity towards C improves at lower concentrations of A and C.

Reactor design, configuration, and parameters influence temperature and concentration profiles.

Reactor Features

Reactor feature offer an option to optimize process productivity and yield. These features mainly differ with respect to

- Mixing and back-mixing properties
- Heating and cooling capabilities
- Addition and removal of educts and products

These reactor features of Figure 46 control the reaction operating parameters such as

- Temperature
- Pressure (gaseous systems)
- Concentrations of educts and products

During initial synthesis of the rector system, the reactor size is of minor interest. A process designer studies the reactor size in more detail during process analysis optimizing residence time in terms of selectivity, conversion, and productivity.

Matching Reactor with Reaction Features

Reactor synthesis aims at finding the best fit of reaction system necessities and reactor design features. In other words, the reactor features (e.g. mixing, back-mixing, feed addition, product removal, heating/cooling etc.) have to establish reaction conditions allowing the best performance.

Figure 47 illustrates this task for a competing, parallel reaction system.

An analysis of the reaction system in Figure 47 leads to different reactor configuration depending on the respective reaction order (e.g. factor a_i and b_i). The reactor configurations feasible for different values of a_i and b_i are summarized in reactor design rule 9 chapter 5 (Smith, 2004).

A single continuous stirred tank reactor (CSTR) possesses lower concentrations of

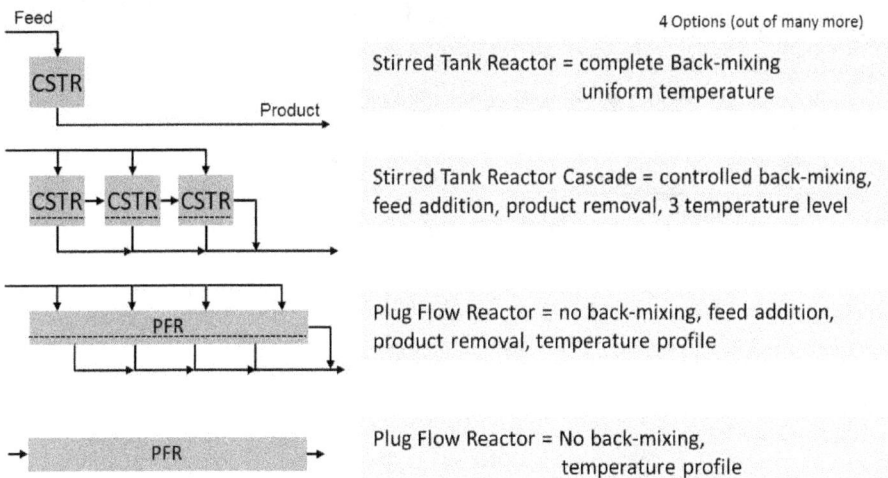

Figure 46: Basic Reactor Configuration

A and B due to back mixing with product. This favors selectivity towards C compared to D in the example of Figure 47.

Absolute rate to generate C, however, is lower due lower concentrations for the CSTR compared to other reactor design.

The selection of the concentration levels in a CSTR represents an optimization problem balancing rate (e.g. reactor investment) with yield (e.g. raw material usage), since the maximal productivity requires a different reactor design and parameters than the maximal selectivity and conversion.

In many cases it is possible to evaluate the reaction-reactor fit qualitatively – do not get lost in detailed calculations too early. Sometimes a sound decision on reactor design and configuration is only possible with quantitative, computational simulation based on measured kinetics.

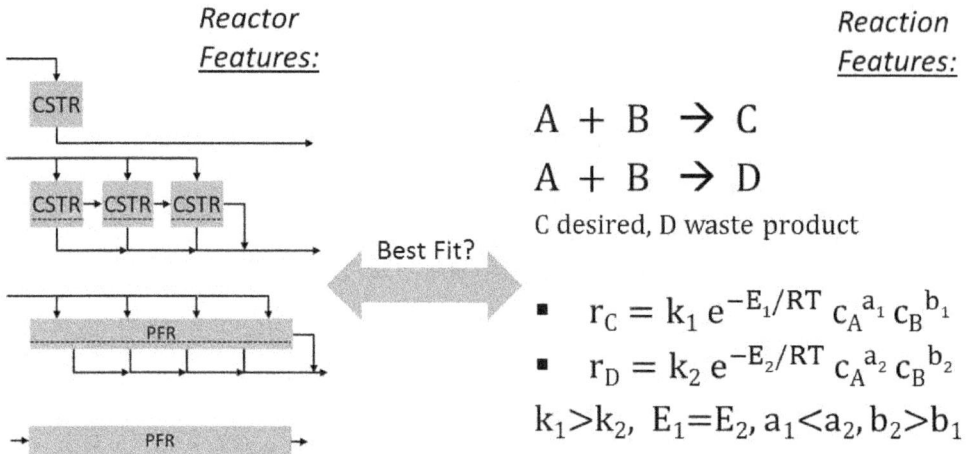

Figure 47: Kinetic versus Reactor Features

Design Case 4 – Reactors

Qualitative selection of an optimal reactor system is demonstrated using the reaction system:

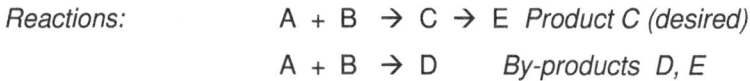

Reactions: A + B → C → E *Product C (desired)*

                      A + B → D   *By-products D, E*

Reaction Rates: $r_C = k_1 e^{-E_1/RT} c_A{}^{a1} c_B{}^{b1} - k_3 e^{-E_3/RT} c_C{}^{c1}$

                      $r_D = k_2 e^{-E_2/RT} c_A{}^{a2} c_B{}^{b2}$

                      $r_E = k_3 e^{-E_3/RT} c_C{}^{c1}$

                      with $E_2 < E_1 < E_3$ and $a_1 = a_2 = c_1$, $b_2 > b_1$

This example consists of competing, parallel, and consecutive reactions. The component C is the desired product. The kinetics of the reaction system is qualitatively described covering activation energies and reaction orders.

The design goal is to maximize the rate towards product C, while minimizing the rates towards E and D.

Based on the levels of the activation energies of the different reactions it is preferable to establish a higher temperatures supporting generation of C compared to D. Unfortunately higher temperatures will also speed up rate to convert C into E. A conclusion could be to start with a higher temperature, but reduce temperature as soon as significant levels of C are generated. This would initially lead to better selectivity C/D at higher productivity, followed by lower temperatures to avoid generation of by-product E.

High concentration of A is acceptable, since selectivity C/D is not affected by concentration of A ($a_1=a_2$), but productivity becomes higher. Low concentration of B is beneficial, since rate towards C relatively increases compared to rate towards D at lower levels of B ($b_2>b_1$).

The rationale outlined leads to the following conclusions:

- High temperature favors generation of C compared to D, but accelerates the reaction of C to E as well. A temperature profile starting with a higher temperature followed by decreasing temperature could support selectivity towards C → Cascade of inexpensive CSTRs desirable.
- Selectivity C improves, if concentration of A is high and concentration of B is low. A gradual addition of B is favorable, while component A is directly added to the reactor → Gradual addition of B to cascade beneficial.
- Low concentration of C avoids generation of E → Steady removal of C beneficial.

A cascade of continuous steered tank reactors provides inexpensive, large residence times, but still different, discrete temperature and concentration levels for each reactor. Figure 48 shows a reactor cascade consisting of three stirred tanks and qualitative concentration and temperature levels.

A membrane reactor creating a plug flow and allowing continuous removal of C could be most optimal provided a separation of the product C from all other components is technically feasible.

The heuristic rules of chapter 5 (level 2) should lead to the same design, since the rules reflect most of the logic applied.

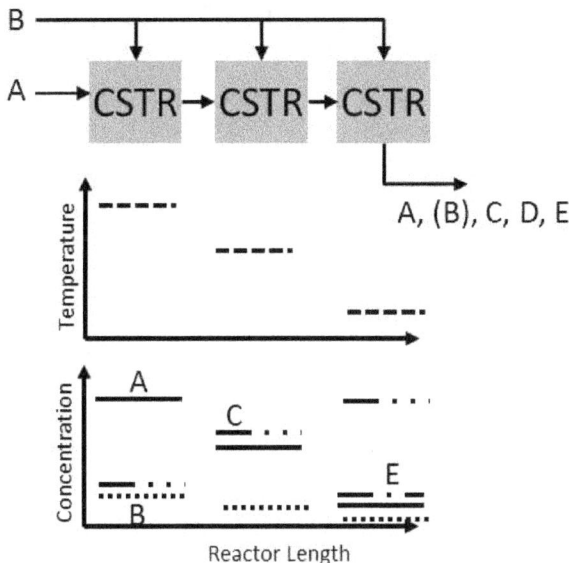

Figure 48: Reactor Design with Concentration Profile

Rule 4 covers the link between activation energy and temperature levels emphasizing that reaction steps with higher activation energies are preferred by higher temperature levels. Rule 9 applies to competing, parallel reactions. Rule 6 recommends a reactor without back mixing for consecutive reaction systems. Rule 12 recommends steady addition of B and removal of C.

Obviously different rules can be applied to a problem resulting in contradictory recommendations. Sometimes the sequence using various rules may also lead to different results.

The optimal number of cascade reactors, temperature profiles, and addition distribution of B has to be verified as soon as quantitative data are available. Of course, a plug flow reactor represents another solution allowing tailor-made temperature and concentration profiles (of B).

Unfortunately, the information on the reaction system is generally not available at the start of process design. A process designer, therefore, has to insist on getting these data from the development scientists.

3.1.3 Design of Recycles and Separations

After designing an initial reactor system, the next step includes the evaluation of separation technologies, separation sequences, and recycles options. The goal is to produce a pure product, separate by-products and recycle unused feed.

<u>Recycle Design</u>

Level 3 of the hierarchical design methodology generates the recycle structures of the new process (Figure 37). The reactor system design delivers in addition to initial reactor type, structure, and operating parameter the composition of the reactor effluent – at least the chemicals and feasible concentrations. The recycle requirement is evaluated to design the separation system.

Figure 49 describes the design task concerning necessary recycles and/or purges schematically. Inexperienced (and very often experienced) designers immediately jump to a detailed evaluation of the separation system. During this design phase, designers need to avoid an early selection of specific technologies. A first preliminary evaluation of the recycle/purge options is recommendable.

This includes

- to determine reactor outlet composition or potential composition range
- to evaluate impact of recycle for each component on conversion, selectivity and productivity
- to decide on recycle

Decisions on recycle and purge streams are made using heuristics, since there are still not sufficient data available generally. Chapter 5 (level 3) summarizes some heuristic rules. Based on these rules a process designer selects an initial recycle scheme that will be used to select separation technologies best suited for the separation task.

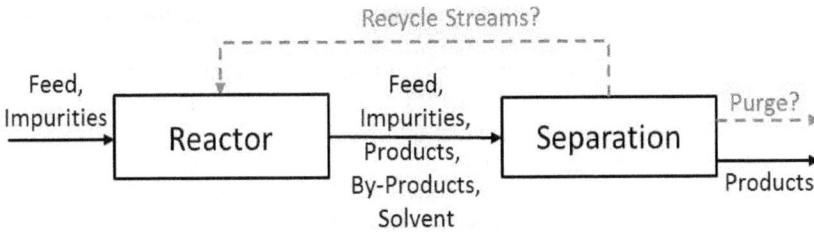

Figure 49: Design Task for Recycles and Purges

A reactor – separation structure with recycle streams and/or purges generally includes an optimization issue. The reactor effluent composition links the reactor system with the separation task. An increase in reactor conversion leads to a reduced requirement for the separation section. Often higher conversion requires larger reactors (e.g. more capital spending).

The reactor design solution proposed in design case 4 is used to illustrate the recycle configuration task.

The effluent of the stirred reactor cascade is likely to contain a mixture of components A, C, D, and E. Recycle of C is not necessary, since C is the desired product. Heuristic recycle design rule 1 of chapter 5 recommends recycling unused feed A (and B, if not totally consumed). A re-cycle of components D and E is detrimental, since these components will accumulate in the process (rule 4).

Since low concentration B is beneficial for the selectivity, an excess of component A leads to a complete consumption of component B in the last reactor avoiding separation of B. Small quantities of B could also be purged with a by-product from the process provided the property data support this concept. Figure 50 shows the recycle scheme.

The recycle scheme includes several optimization issues:

▪ A purge of remaining B instead of a recycle of B makes only economic sense,

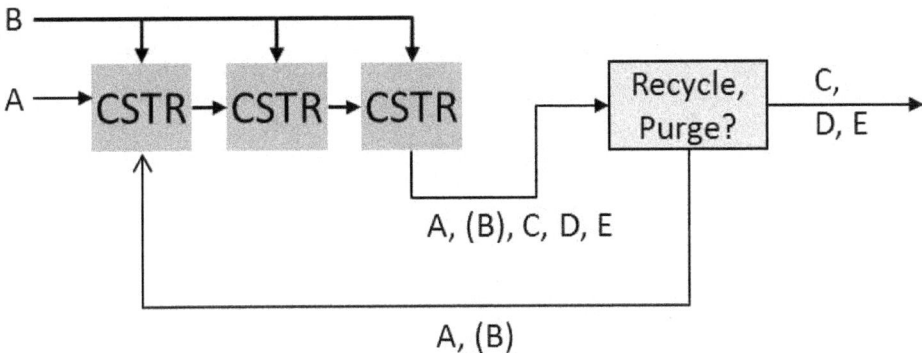

Figure 50: Recycle / Purge Scheme

if the component B is not very expensive or – from a sustainability point of view – not toxic or limited.

- Larger reactor volumes lead to a smaller amount of unreacted feed to be recycled. Smaller reactors that result in more A and potentially B require bigger separation units and probably more energy.
- The design and operation of the reactor system requires an optimization of yield (selectivity) and rate (productivity), since higher temperature improves productivity, but generates more by-products as well.
- Lower temperature will improve yield. This benefit needs larger reactors due to lower productivity. An optimization evaluation has to study the trade-off between capital investment (and some operating cost) and raw material expenses.

During initial process design, the primary goal is to evaluate whether there is an optimization issue later dealt with in process analysis (chapter 3.2.1).

Design Case 5 – Recycle Design

A special situation can occur if an equilibrium reaction is part of the overall reaction system. The reaction system of design case 4 – reactors is slightly modified leading to a by-product D.

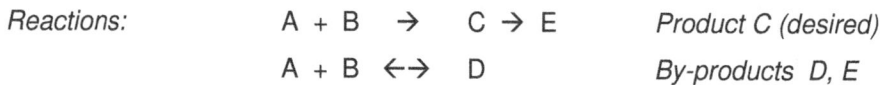

Reactions: A + B →   C → E   *Product C (desired)*

A + B ←→   D   *By-products D, E*

Reaction Rates: $r_C = k_1 e^{-E_1/RT} c_A{}^{a1} c_B{}^{b1} - k_3 e^{-E_3/RT} c_C{}^{c1}$

$K_D = constant$

$r_E = k_3 e^{-E_3/RT} c_C{}^{c1}$

with $E_2 < E_1 < E_3$ and a1=a2, c1, b2>b1

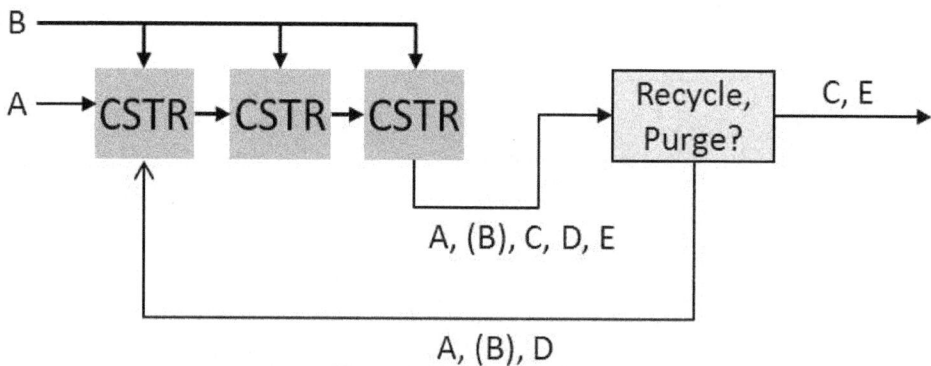

Figure 51: Beneficial Recycle of Equilibrium By-Product D

Does this influence the selection of recycles as shown in Figure 50? Since the by-product D is formed in an equilibrium reaction, a recycle of by-product D to the reactor will convert D back into A and B according to the equilibrium. The newly formed A and B will partially be converted into the desired product C. When the by-product D is always recycled to the reactor, the by-product D can be recycled to extinction. Consequently, the process shown in Figure 51 produces only product C (and by-product E).

Recycling of undesired by-products is only beneficial, if the by-product can be converted in feed materials. A case-by-case evaluation is always necessary.

Design of Separation Technologies and Sequences

The effluent of the reaction system contains various components that need to be recycled to the reactor, purged as waste, or processed as product. Based on the recycle structure the separation design includes the selection of separation technologies and separation sequence (Figure 52).

Insufficient data often make the design task difficult. Often this still allows a proper design, but sometimes the design of the process separation section needs to be refined during later stages of the process design procedure.

The separation system design includes the following tasks:

- Define separation sequence
- Select separation unit operations
- Evaluate impact of separation on recycle and reaction

There are two tasks to solve – select separation technologies suited for each separation and configure the best sequence for the separations system. Both tasks are strongly linked and may be repeated several times to find the best solution.

Figure 53 gives a summary of the most relevant separation. Each of the separation technologies applicable to different separation areas possesses advantages and disadvantages. There is a huge variety of literature on separation design – sequence and equipment – available (Smith, 2004).

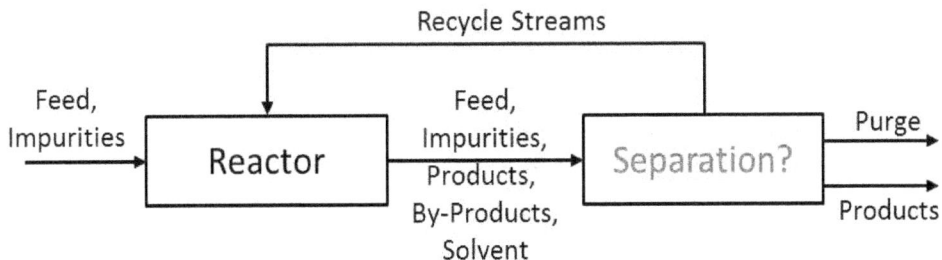

Figure 52: Separation Design Task

- **Multiphase Separation**
  - Gas-liquid
  - Gas-solid
  - Liquid-liquid
  - Liquid-solid
- **Solid Separation**
  - Gravitational Settling
  - Centrifugal Separation
  - Sieving / Filtration / Membranes

- **Liquid Separation**
  - Evaporation, Distillation
  - Chromatography, Membranes
  - Distillation
  - Extraction
  - Ad-/Absorption/Chromatography
  - Crystallization
  - Drying / Granulation
  - Extrusion

Figure 53: Separation Technologies

Due to incomplete data, the task to find a structure cannot be accomplished straightforward applying a mathematical algorithm. Once more, a process designer creates an initial separation design based on heuristic rules. As always with heuristic rules, these rules may be ambiguous or even contradict each other – trial and error is unavoidable. Promising designs are then simulated and optimized.

An overview on various heuristics concerning separation design is given in chapter 5 (level 4a+b). The heuristics give a guideline to select the technology best suited and some indications where to position the technology within a sequence. A special focus is given on design rules for distillation sequences (level 4b).

Based on the composition of the reactor effluent, various feasible separation technologies are screened with these heuristic rules to establish a list of technologies suited for the various separation tasks. Subsequently a sequence of separations is chosen once more by applying the heuristic rules.

Some data on material properties are added to the design case 4 – recycle.

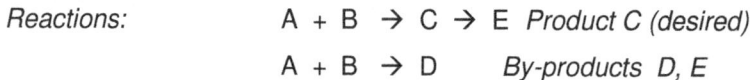

Reactions: $A + B \rightarrow C \rightarrow E$ *Product C (desired)*

$A + B \rightarrow D$ *By-products D, E*

Reaction Rates: $r_C = k_1 \, e^{-E_1/RT} \, c_A{}^{a1} \, c_B{}^{b1} - k_3 \, e^{-E_3/RT} \, c_C{}^{c1}$

$r_D = k_2 \, e^{-E_2/RT} \, c_A{}^{a2} \, c_B{}^{b2}$

$r_E = k_3 \, e^{-E_3/RT} \, c_C{}^{c1}$

with $E_2 < E_1 < E_3$ and a1=a2, c1, b2>b1

Boiling Point at ambient air pressure:

$T_A < T_E \ll T_B \ll T_C < T_D$ all liquids

$A : B : C : D : E = 30 : 2 : 8 : 10 : 50$ in quant. %

Component B can economically be purged with one of the by-products.

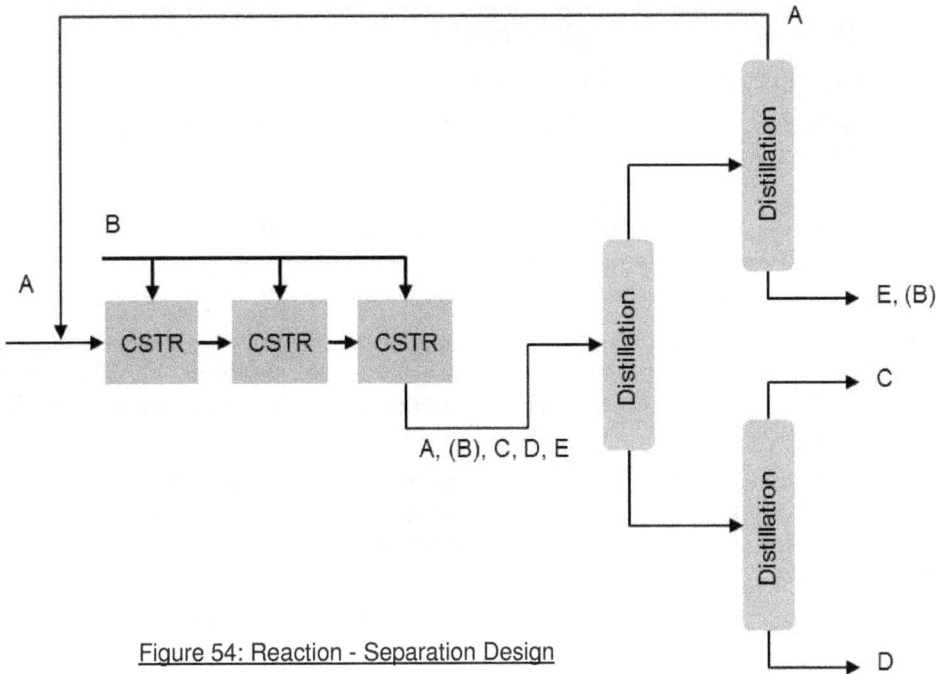

Figure 54: Reaction - Separation Design

Since all of the components of the reactor are liquid at ambient air pressure, separation rule 1 of chapter 5 (level 4b) suggests distillation as separation technology of choice for all separations.

The first distillation separation is an "equal" split of C and D from A, B and E as recommended by heuristic rule 4. Subsequently the mixture C and D is separated generating C as pure top (heuristic rule 7). In a parallel distillation, the component A is separated from E containing small quantities of B. Stream E acts as purge for B. Figure 54 shows the resulting flow sheet.

Design Case 6 – Separation design

Finally, an example is discussed including separation techniques beyond distillation.

This application to select the separation technologies and sequence is demonstrated for a mixture of five components A, B, C, D, and E that is to be separated into the 4 products. Relative quantity, phase state (solid, liquid), boiling temperature and special features characterize these products:

Component:		Quantities:
A:	Small molecule, liquid, boiling point T_A	47%
B:	Solid particles	3%
C:	Small molecule, liquid, boiling point T_C	20%

D: Larger molecule, damaged by higher temperature T_D 10%

E: Small molecule, liquid, boiling point T_E 20%

Relative boiling points: $T_A < T_C < T_E <<< T_D <<<<<< T_B$.

There are four technologies available to separate the mixture. Figure 55 describes these technologies:

-   Sieving Filtration for solids
-   Membrane Separator for solids and large molecules
-   Chromatographic Absorber for small and large molecules
-   Distillation Column for molecules

Based on the information on the components to be separated and the available separation techniques, heuristics provide an initial selection on applicable separation technologies and a feasible sequence for each component A, B, C, D, and E.

Sieving filtration capable to handle solids becomes the choice to remove the solid component B from the mixture. Membranes provide an alternative technology for this task, but are probably more sensitive to damage, more energy consuming and more expensive than sieving filtration.

Heat sensitive component D can be isolated applying either membranes or chromatography (level 4a/rule 11, 15). Chromatography is generally more specific then membranes depending on separation task. A decision between membrane or chromatography technology requires more information on the properties of

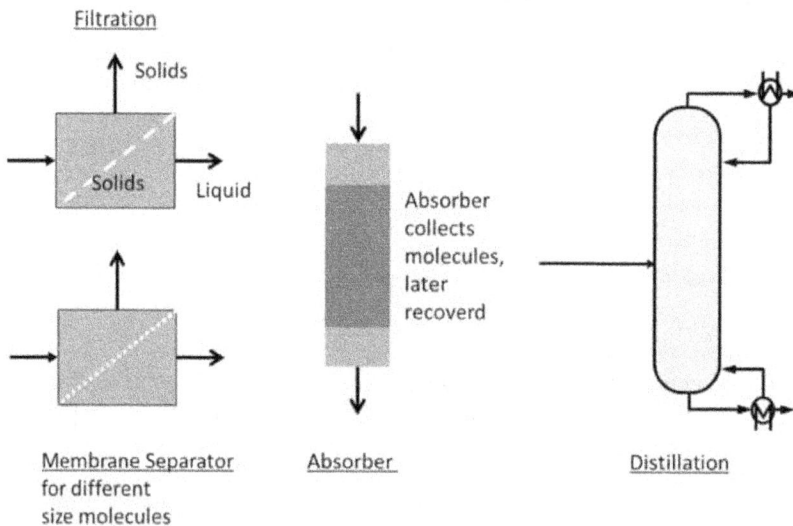

Filtration

Solids

Solids   Liquid

Absorber collects molecules, later recoverd

Membrane Separator
for different
size molecules

Absorber

Distillation

Figure 55: Separation Technologies

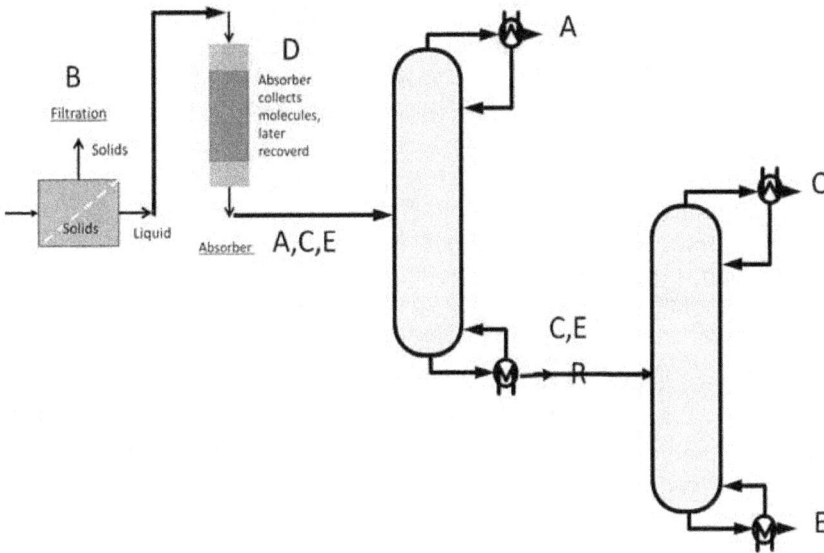

Figure 56: Design of Separation System

component D:

The liquid components A, C, and E are liquid at ambient air pressure. Distillation represents the standard separation technology applied first (level 4a/rule 2).

After selecting feasible separation technologies, the process designer addresses the separation sequence next.

Solid B is removed first (level 4a/rule 3) to avoid plugging of other separation equipment.

Component D is removed as second step (level 4a/rule 11, 12) before temperature is increased to unacceptable levels.

The liquid component A is removed in third position (level 4b/rule 3, 4). The component C and E are separated last.

Figure 56 shows the resulting separation sequence.

Generally, the design of a separation system occurs in a trial and error mode. The system design is continuously reevaluated when more process data become available.

In case of complex, difficult questions, involvement of company-internal or external experts is always recommendable – particularly to verify a technology / sequence choice.

3.2 Process Analysis – It's about Simulation, Targets, and Benchmarks

Process synthesis generates an initial structure for a new process design. Process parameters are only considered as far as they are relevant to find the optimal structure. Process analysis studies and optimizes these process parameters.

Process analysis covers all activities to model, simulate and optimize process parameters for a given process structure. Modeling transfers available process knowledge into a mathematical model. This model is then used to simulate the impact of an individual set of process parameters on the process performance – productivity, yield, and profitability. The model is also used to optimize the parameter data set leading to an optimum value for a given function such as a maximal yield or minimal operational cost.

Modeling, simulation, and optimization of chemical process parameters represent the main tool of process analysis. In addition to classical analysis, targeting and benchmarking represent two other tools indispensable during process analysis.

3.2.1 Simulation – The Core of Analysis

Modeling, simulation, and optimization are mature areas in chemical engineering. Models of most chemical unit operations have significantly advanced during the last 50 years. Algorithms for process flow sheeting and optimization are mathematically robust and easy to be used. Process simulation programs are generally part of engineering suites integrating data and knowledge management from development through engineering until construction. Chapter 2.2 already discussed general features of simulation software briefly.

Modeling, simulation, and optimization of chemical processes represent the core of many chemical engineering courses at universities. Many excellent chemical engineering books give comprehensive overviews on concepts and tools (Seider, Seader, & Lewin, 2004) (Smith, 2004) (Sundermann, Kienle, & Seidel-Morgenstern, 2005). This chapter only discusses two aspects of process analysis: Process simulation to generate process insights and process parameter optimization.

Simulation for Process Development

Process models mainly consist of process mass balances, energy balances, kinetic data, material property data, and economic descriptions as well. These models can be steady state or dynamic models.

Computational modeling, simulation, and optimization are particularly beneficial when accompanied by experimental studies. Experimental studies provide data on kinetics or material properties or verify insights generated by computational simulation. An appropriate design of experiments is a prerequisite to support simulation optimally.

Buchaly, Kreis, and Gorak demonstrated the approach combining computational simulation with experimental studies for a heterogeneous catalytic esterification of propanol and propionic acid to generate propyl propionate.[9]

Since esterification represents an equilibrium reaction, continuous removal of the product from and immediate recycle of the educts to the reaction favor reaction performance. Figure 57 illustrates an integrated process combining a reaction column and a pervaporation unit. Chapter 3.4 discusses the design method to create such an integrated process in detail.

The reaction column performs the catalytic reaction in the reaction section. Propionic acid (A) is fed to the column above the catalytic section, while propanol (B) is added to column below this section. Both educts react to water (D) and propyl propionate (C). Propanol and water climb up in the column, while propionic acid and propionate move down the column due to their boiling features.

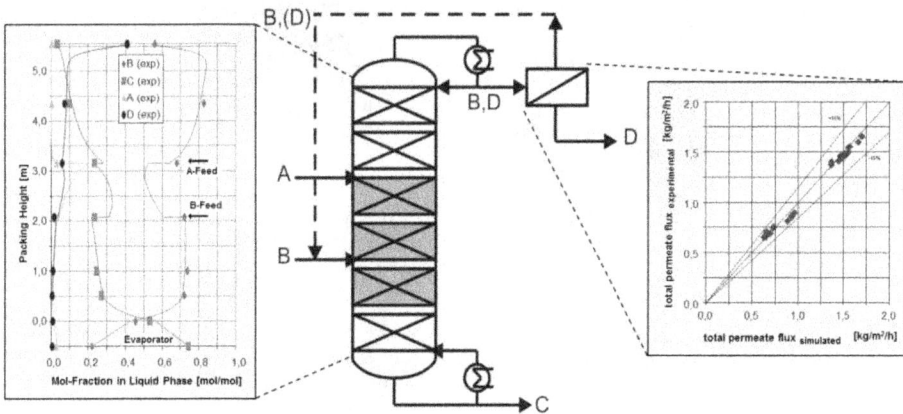

Propionic Acid (A) + Propanol (B) ←→ Propyl Propionate (C) + Water (D)

Figure 57: Hybrid Process Integration (Source: TU Dortmund)

The lower part of the reaction column is designed to separate propyl propionate from propionic acid. Propyl propionate is removed from the column as bottom product.

Propanol and water form an azeotropic mixture that cannot be separated by classical distillation. Pervaporation is used to separate this mixture and to withdraw water from the process. Propanol is returned to the reaction column.

Propanol and propionic acid are continuously recycled to the reaction column where they are converted into the desired product.

---

[9] Buchaly, C., Kreis, P., & Gorak, A. (2008, 1+2). CIT.

A reaction column is an elegant example how to utilize the equilibrium in an advantageous way, provided the properties of educts and products allow performing reaction and separation simultaneously.

The optimal design of a reaction column requires experimental studies and computational simulation as well. Only experimental _and_ computational evaluation of the reaction column can deliver a comprehensive insight into the complex technical system.

The experimental laboratory reaction column is shown in Figure 58. The column consists of reaction section, rectification, and stripping zone. Catalyst covers the surface of the packing in the reaction section. A pervaporation unit separates the

Figure 58: Pilot Plant (Source: TU Dortmund)

azeotropic mixture. Core device is a plate module containing the membranes. Several sensors measure local temperatures and component concentrations.

Kinetic data for the equilibrium reaction and material property data for separation form the basis of the process model. This model simulates the process linking the reaction column and membrane unit. Figure 57 describes the calculated and measured distribution of propanol, propionic acid, water, and propyl propionate in the reaction column on the left. Figure 57 also shows a comparison of calculated and experimental data on the permeate flux (on the right).

As soon as experiments validate the process model, the new model becomes a powerful tool to generate new insights into the process or to optimize process parameters.

Process simulation is the main design tool for process optimization. Process models can be expanded to cover additional aspects – for example dynamic performance during start-up operations or safety issues due to critical process conditions. Detailed engineering uses process models to specify equipment, design process layout, or prepare cost estimates.

Simulation for Process Optimization

Process simulation is particularly beneficial to optimize process parameters as soon as an initial process structure was established. Process optimization relies on computational models of simulators.

The design case 1 introduced earlier demonstrates a standard optimization tasks during process analysis.

Figure 59 describes the impact of product effluent concentration on reactor/separation equipment size (e.g. capital cost) and operating expenses (e.g. separation energy consumption).

Higher product concentration of the reactor cascade effluent requires more or larger reactors resulting in a bigger reactor investment. Higher product concentration, however, causes a decrease of separation cost. These contrary trends result in an optimal effluent product concentration characterized through minimal annual manufacturing cost.

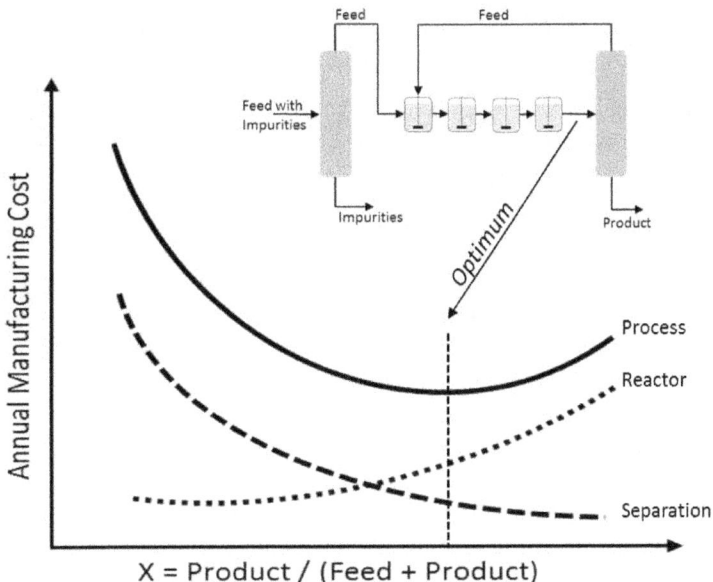

Figure 59: Recycle Optimization

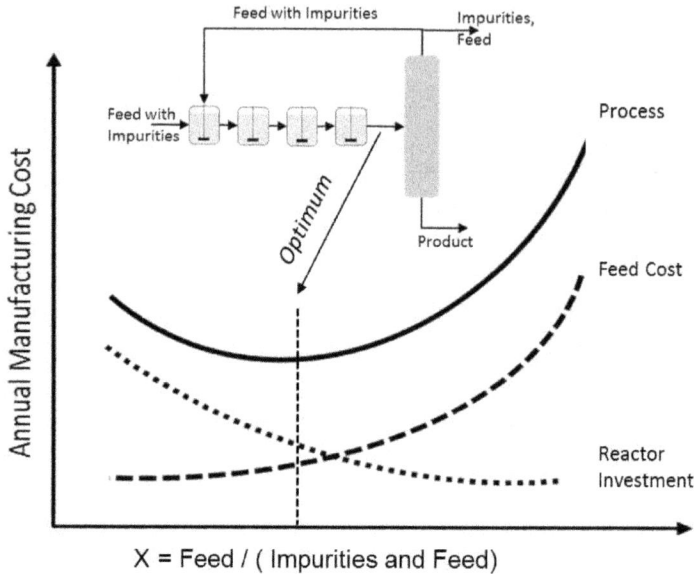

X = Feed / (Impurities and Feed)

Figure 60: Purge Optimization

In Figure 59, a distillation column separates the impurities from the feed before the reactor cascade. A purge stream to remove impurities represents an alternate process design avoiding the column to remove impurities (Figure 60). For this design, a cost trade-off exists between increased feed cost due to feed losses in the purge stream and cost savings due to avoided distillation column investment and energy consumption. The quantity of the purge stream defines the impurity level in the process – larger purge stream, lower impurity level, but bigger feed losses.

The optimization problem in Figure 60 is similar to the earlier example including one additional parameter (e.g. purge stream). A comparison of the annual cost for both structures (Figure 59 and Figure 60) provides the optimal structure and parameters. This quantitative comparison may correct the result from the qualitative, heuristic evaluation of chapter 2.2.

Finally, Figure 61 studies two alternate structures. Impurities are removed before the reaction section, since impurities could react with product at higher temperatures (upper right). The alternate structure show in Figure 61 (lower part) was rejected because impurities might react with product in a separation column at higher temperature.

The impurity removing column is significantly bigger, if the impurities are removed upstream of the reactor cascade (Figure 61 – upper part), since this distillation column has to evaporate the complete feed volume.

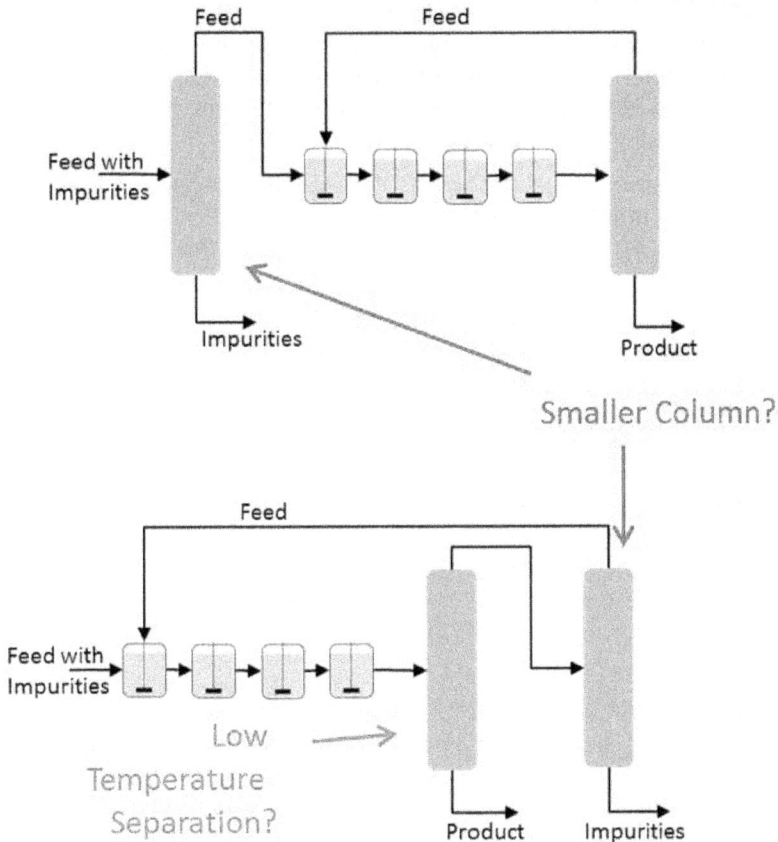

Figure 61: Equipment Optimization

From an investment and operation viewpoint, an impurity removing separation positioned downstream of the reactors is smaller and cheaper than a impurity removing column upstream of the reaction section, because most of the feed was already converted into product and removed from the process.

Assuming an extraction process is available operating at low temperatures, a process design consisting of two separation systems after the reactor cascade becomes a viable alternative. A quantitative evaluation of all three designs (upstream or downstream impurity columns or purge) operated at the optimal process parameters leads to the best design. In many cases, very rough estimates may already allow a decision on the structure. Nevertheless, this simple example emphasizes the complexity of process design requiring a flexible process designer open to revise earlier decisions.

Are there tools available to guide the designer through the design jungle?

3.2.2 Targets – Guiding the Design Progress

Since there is no tool available to synthesize the optimal process in an uncertain, complex environment straightforwardly, a tool is desirable to guide and compare the actual process design progress with a target characterizing the performance of the optimal, but yet unknown process.

The fact that some chemical processes have continuously been improved without introducing a disruptive new technology indicates that earlier process versions have not been the optimum despite all declarations that this is now the best design.

Targets are necessary to guide the activities to design a new process. Targets have to quantify the optimal performance of a process. These targets should be derived from first principles. They must not be derived using the actual design methodology to avoid "self-deception".

Figure 62 gives a schematic description of the potential target framework. In process design, product quantity is specified as a goal, while all other input and output in Figure 62 are considered possible targets.

For example, if a designer knows the minimum energy consumption target of a process, he could compare the energy consumption of his actual design with this target. As soon as the energy consumption of the actual design is close to target, the designer quits his design effort, even if a slightly better design would be possible.

Targets are sometimes confused with process drivers. Process drivers are parameters that strongly affect process performance. A focus on process drivers is

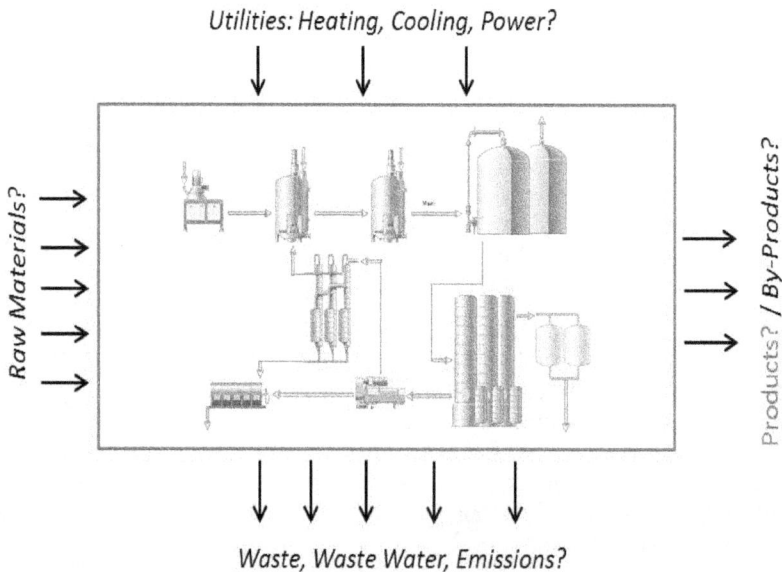

Figure 62: Process Targeting

beneficial to waste no effort on irrelevant parameters. In contrast to process drivers, process targets form a benchmark used to compare a design solution with the target optimum.

As soon as a design's performance is close to a process target, further process optimization possesses only limited additional potential. Even if a slightly better design exists, the close-to-target design is likely to remain competitive in the market. Process drivers and process targets supplement each other during the design effort.

Targeting deals with the technical and economic performance of a process design. Technical targets cover the optimal technical performance.

There are four major *Technical targets*:

The *Material Utilization Target* defines the minimal raw material usage to produce a product (kg raw materials used per kg product produced).

The *Energy Utilization Target* defines the minimum energy input to produce a product (kWh energy consumed per kg product produced).

The Operating Materials Utilization Target defines the minimum operating materials used to manufacture a product (kg operating materials per kg product). Water usage could be a special target measured in kg fresh water per kg product.

The Sustainability Target defines the minimum environment impact. A carbon footprint could serve as target measured in kg CO_2 equivalent per kg product.

Technical targets are transformed into *Economic Targets* when cost and revenues are taken into account. The economic target for the material usage follows from the technical target as:

$\Delta ET(Materials) = \sum Target\ Usage_i * Price_i / Unit\ Usage_i$ with material i=1..n

A comparison of an economic target with an actual economic performance leads to an economic potential. The "economic" difference between actual raw material usage and target usage describes an economic potential for better raw material utilization:

$EP(Materials) = \sum Actual\ Usage - Target\ Usage)_i * Price_i / Unit\ Usage_i$

The economic potential EP actually quantifies the gap between the optimal and the actual process performance.

Technical and economic targets are calculated for processes, plants, or even total supply chains. Targets and potentials are often defined as specific targets or potentials.

For example, energy consumption is generally a major technical target (kW/kg product) linked to an economic target (€ for energy/kg product).

Figure 63 summarizes the most relevant technical and economic targets and respective methodologies.

Target:	Technical:	Economic:	Targeting Method:
Raw Material Use	Yield $_{maximum}$	Material Cost	Mass balances, Stoichiometry
Energy Consumption	Energy $_{minimum}$	Utility Cost	Pinch Point Methodology
Water/Air	Use $_{minimum}$	Waste Cost	Pinch Point Methodology
Waste	Volume $_{minimum}$	Waste Cost	Mass balances, Stoichiometry
Environment	Impact $_{minimum}$	Certificate Cost	Carbon Foot Print

Figure 63: Technical and economic Targets

One feature of technical and economic targets needs to be emphasized with respect to the design of processes, plants, or supply chains:

The different targets cannot be achieved simultaneously. Minimum raw material usage may not coincide with minimal energy or capital or labor usage. This is true for specific economic targets for energy, raw materials, etc. as well.

While technical targets are always specific targets that cannot be added up, economic targets use the same currency and add up to an economic potential for a process, a complete plant, or an overall supply chain. The economic potential can be used to search for the optimal process, plant, or supply chain configuration.

This is also valid for sustainability. Although it is rather challenging, the sustainability performance needs to be translated into financial parameter as well, since decision-making is ultimately done, based on profitability.

Process designers always try to achieve a technical performance close to the technical targets (or better his favorite technical target) and sometimes badly fail when the overall economic process potential is valued. For these reasons, targets and potentials cannot be applied in process design rigorously. They need to be used flexible and adapted to the task.

The following chapters illustrate technical targets, capital targets, economic targets, and economic potentials for process design.

Energy Utilization Targets (Pinch Point Methodology)

The *Pinch Point Methodology* is a well-known, established concept to develop energy targets for new processes and retrofitting as well. There are excellent textbooks on the pinch point methodology available to further deal with this methodology in detail (Smith, 2004), (Klemes, 2011), (El-Halwagi, 2012). Only the basic idea of the pinch point methodology is illustrated in this chapter.

The energy consumption of a process is generally minimized by using the heat generated in one process unit to supply heat to other units. Heat recovery networks

Heating requirement
without heat recovery:
5700 kW

Heating: 2500 kW

160°C

Cooling requirement
without heat recovery:
5500 kW

210°C

50°C   210°C   270°C   160°C

Reactor

Separator

Heating: Cooling:
3200 kW   1980 kW

Heat load:
$$\Delta H = c_e * V * \Delta T$$
with c_e heat capacity
        V flow rate
        ΔT temperature difference

60°C

220°C

Cooling: 3520 kW

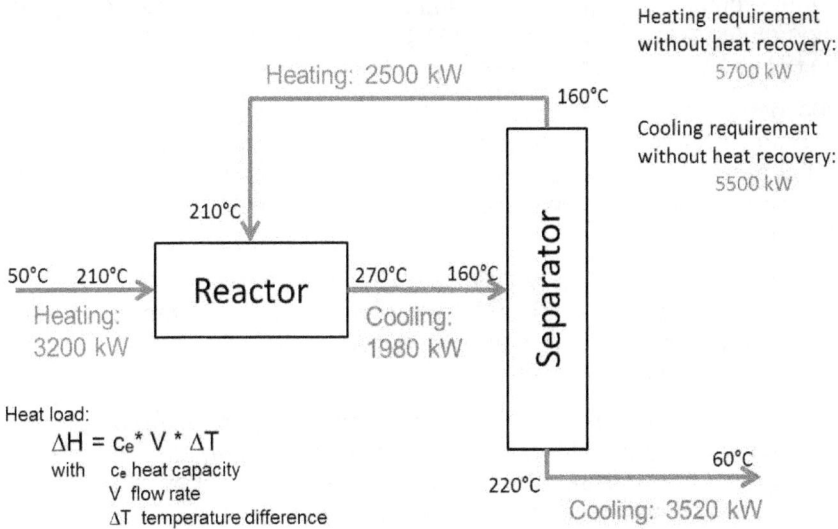

Figure 65: Process Flowsheet for Pinch Analysis

are implemented to distribute heat within a process optimally. The pinch point methodology allows estimating the minimal energy consumption of a process design.

Figure 65 describes a simple flow sheet. This consists of a reactor and separator with two cold streams requiring heating and two hot streams requiring cooling. This process requires 5700 units heating and 5500 units cooling in case there is no heat recovery implemented. Targeting for such a process should provide a minimal energy supply target before a heat recovery system is designed.

As a first step, the four cold and hot streams of the process are described in a

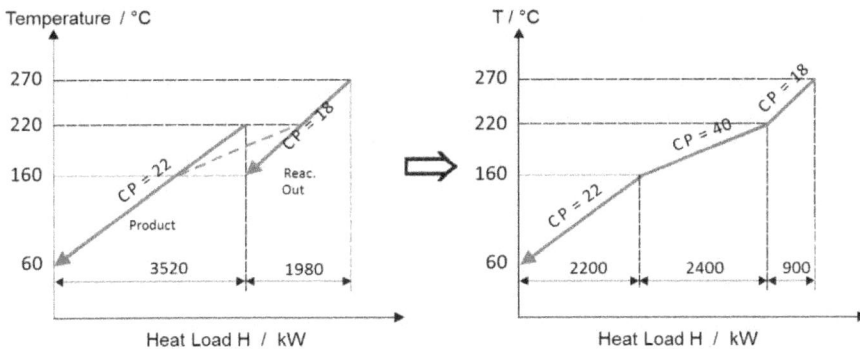

Temperature / °C

270
220
160

CP = 22
CP = 18
Reac.
Out

Product

60
          3520   1980

Heat Load H / kW

T / °C

270
220
160

CP = 18
CP = 40
CP = 22

60
          2200   2400   900

Heat Load H / kW

Figure 64: The Composite Curve for Hot Streams

temperature – enthalpy diagram containing all the information about heat quantities and temperature levels. The cold streams can be combined into one cold curves, while the hot streams are summarized in a hot curve.

Figure 64 demonstrates the procedure for the hot streams. This composite hot stream contains the information about the enthalpy and temperate level of the process hot streams. In a complex process, such a curve can include hundreds of streams. The composite curve condensates the enthalpy and temperature information into one curve.

The core element of the pinch point methodology is summarized in the so-called composite curve diagram as shown in Figure 66. The diagram illustrates the information on heat and cooling demands in two curves. The process literately is "converted" into a "huge" heat exchanger.

Heat can be used to heat the cold streams when heat is transferred from the hot composite curve to the cold composite curve as shown in Figure 66. The composite curves also describe the heating demand (called hot utility target), the internally transferred heat, and the cooling requirement (called cold utility target) of the process in Figure 65 without designing a heat recovery network.

Figure 66 provides insight into the key feature of a chemical process. The process can be divided into heat sources with excess heat and heat sinks with heat deficit. The heat sink is located above the pinch, while the heat source is found below the pinch.

If heat is transferred from above the pinch to areas below the pinch, heat is actually removed from the heat sink (heat deficit) and moved to areas with already excess heat (Figure 67). The heat transferred across the pinch cannot be replaced with

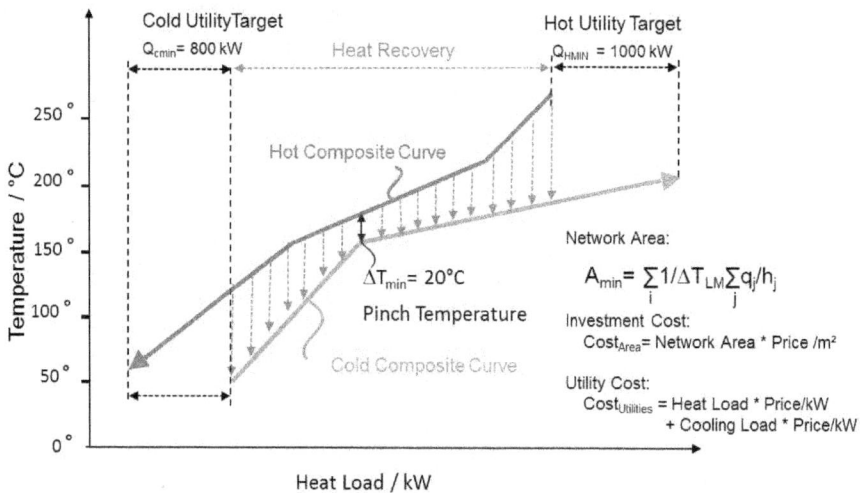

Figure 66: Pinch Diagram with Composite Curves

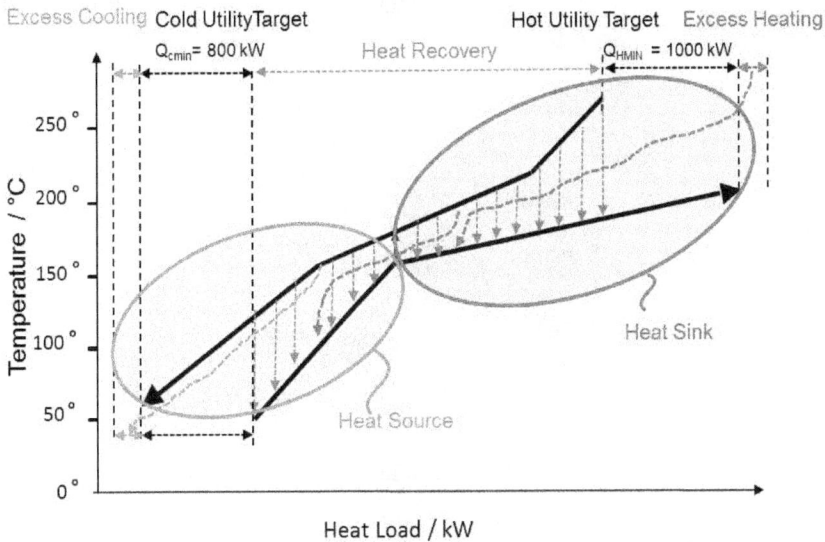

Figure 67: Pinch Point Rule – No Heat Transfer across Pinch

other process heat, but needs to be added with excess heating.

The pinch rule simply states that heat must not be transferred across the pinch. A transfer will always result in additional heating and cooling requirements. As shown in Figure 67.

Important conclusions results from this insight:

- Do not transfer heat from the heat sink to the heat source area, because this transfer deteriorates the energy imbalance of the process.
- Do not add heat generating operations to the heat source area or heat using operations to the heat sink areas.

Design Case 7 – Reactor and Distillation Integration

Let us assume that the reactor and distillation scheme require or provide heat as illustrated in Figure 68. The exothermic reactor requires cooling of 500 kW to keep the temperature below 270°C. The evaporator and condenser of the distillation column require or supply energy in the range of 1000 kW and 800 kW.

This information translates into a pinch point diagram as shown in Figure 69. Compared to Figure 66 the overall process heat requirement only increases by 500 kW, although the distillation column requires an additional heating load of 1000 kW. Similarly, the cooling requirement only increases by 800 kW, although cooling of the reactor and the condenser require 1300 kW.

Since the area above the pinch is a heat sink, a heat source like an exothermic reactor adds heat to the system and, therefore, reduces the heating load directly. The distillation column actually transfers heat from the evaporator across the pinch to   the

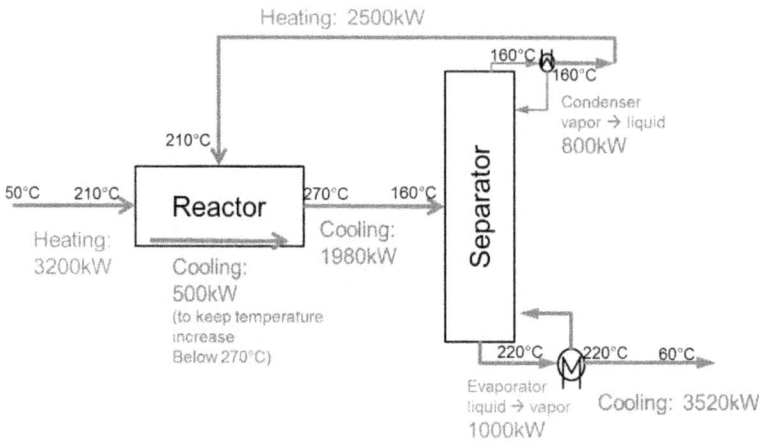

Figure 68: Scheme with Reactor and Distillation

condenser adding even more heating load to the sink and cooling load to the source (Figure 69).

From an energy consumption point of view, it is necessary to integrate the reactor, but not beneficial to integrate the distillation evaporator and condenser.

From an operational point of view, linking unit operations always creates difficulties to operate processes. Therefore, the distillation column performs preferably as a stand-alone unit, while the reactor has to be integrated into the process to reduce the heat demand.

The pinch point methodology explains why the integration of the cooling and heating requirements of reactor and distillation led to this result.

Does the pinch analysis also provide recommendations how to integrate either

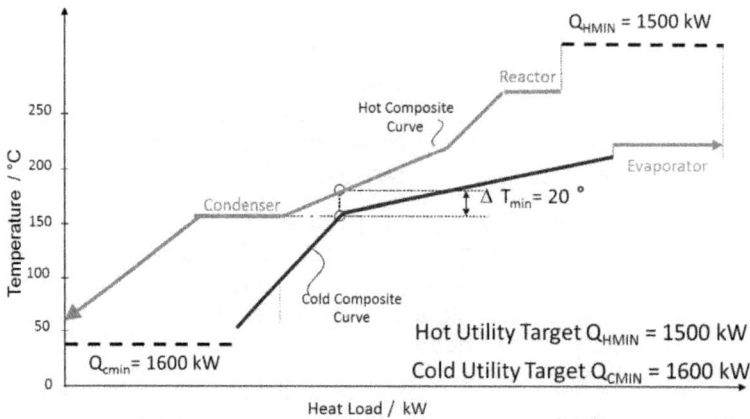

Figure 69: Pinch Diagram with Reactor and Distillation

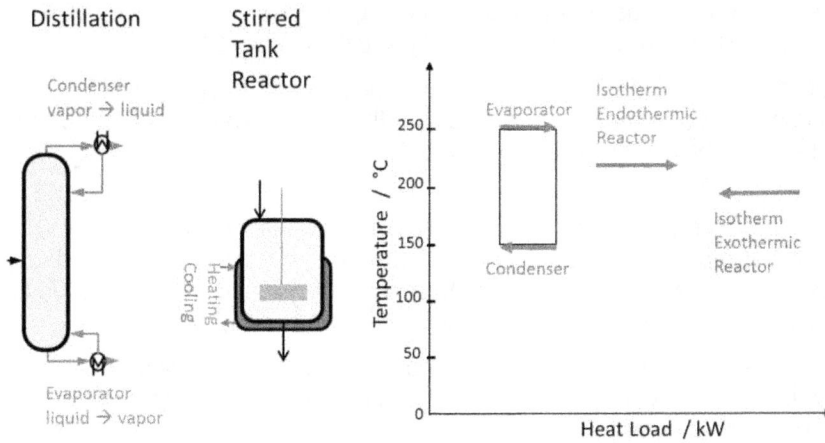

Figure 70: Reactor/Distillation-Description in Composite Curves

reactor or distillation column into an overall process scheme?

Figure 70 illustrates how reactors and distillation columns are simplified described in a pinch point diagram.

Figure 71 demonstrates the impact of reactor integration for exothermic or endothermic, isotherm reactors at different temperature levels on the energetic

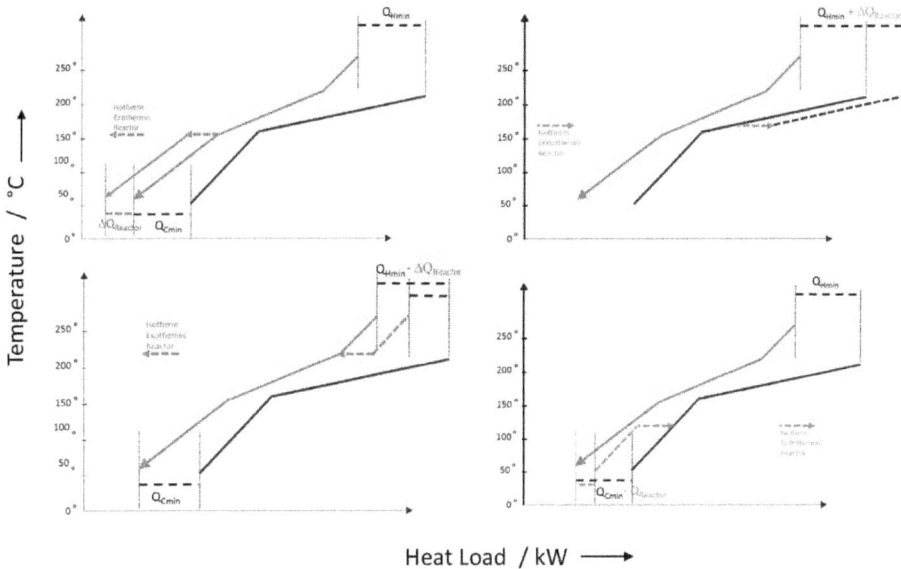

Figure 71: Integration of Reactors into Processes

process performance using the composite curves. The composite curves of this process (e.g. without reactors) was already shown in Figure 66.

First, an exothermic reactor is considered for process integration. Integrating an exothermic reactor above the pinch reduces the heating requirements due to reuse of the reactor heat in the process without adding to the cooling load (lower left case).

The heat requirement of the endothermic reactor can be met without additional energy consumption, provided the endothermic reactor is integrated into the process below the pinch (lower right case). Excess process heat drives the endothermic reactor reducing the demand for cooling capacity.

The "opposite" outcome occurs, if an endothermic reactor is added above or an exothermic reactor is added below the pinch temperature. The heat sink/source logic explains these results.

This logic provides rules on reactor integration:

Exothermic reactors are to be integrated into a process above the pinch temperature. Endothermic reactors are beneficial below the pinch temperature. This integration approach reduces the heat and cooling requirements of a process design.

Figure 72 illustrates the impact of distillation column integration into a chemical process. Figure 66Figure 67 described the composite curves of this process (e.g.

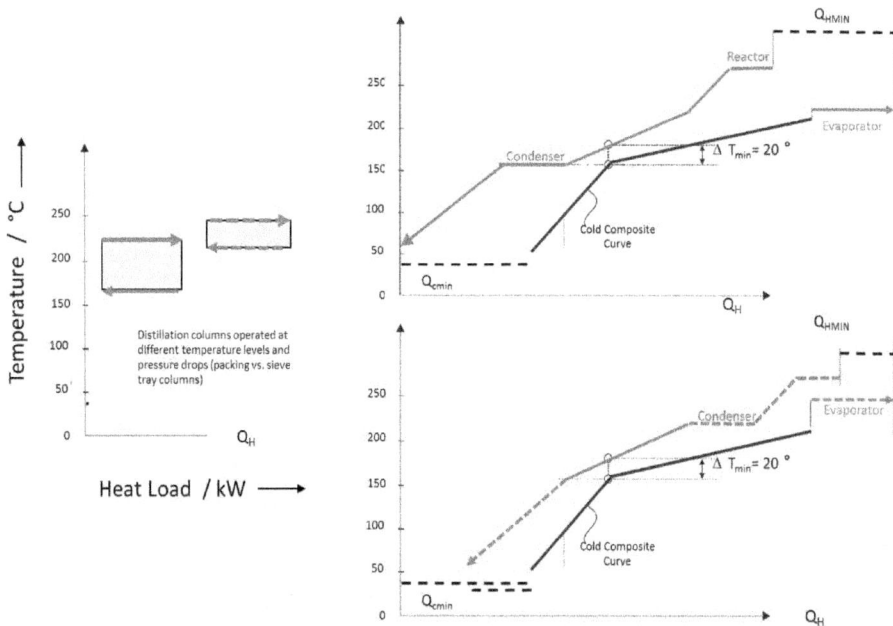

Figure 72: Integration of Distillation Columns into Processes

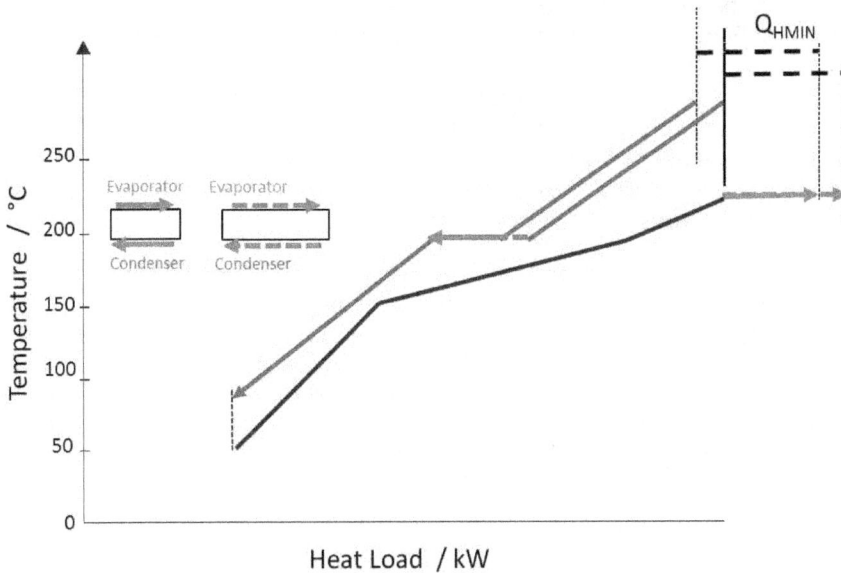

Figure 73: Reflux Ratios and Heat Load of integrated Columns

without distillation columns). There are two different distillation columns shown. If the distillation column with the higher temperature levels is integrated into process, no additional energy is used for the process with column. The heat balance is neutral, since the evaporator heat transferred to the condenser within the column is reused above the pinch to fulfill other operations requiring heat (lower right).

If the column with the lower temperature levels is added to the pinch diagram, heat and cooling requirements of the overall process increase according to evaporator heat load and condenser cooling demand. Heat is actually transferred across the pinch (upper right case in Figure 72).

Distillation columns can only beneficially be integrated into a process network, if the evaporator and condenser are either above or below the pinch point. Only this approach does not influence the energy consumption negatively.

This result can be used to try to change the temperature levels in a distillation column allowing an energy positive integration of the distillation operation. In Figure 72, operation of the distillation column at higher pressure will increase the temperature of the evaporator and condenser as well. Sometimes distillation temperatures could be moved above the pinch temperature. Furthermore, lowering the column pressure may enable a distillation operation only below the pinch.

In addition, replacing a sieve tray design with a low-pressure drop packing will reduce the pressure drop in the column. This measure facilities shifting the temperature levels in a column completely below or above the pinch temperature resulting in an optimally integrated process (e.g. the distillation column is completely integrated above or below the pinch).

Generally, an increase of the reflux ratio is linked to higher energy consumption of this column. Figure 73 reflects this increased heat consumption. Although the distillation column requires more energy, an optimally integrated column does not lead to an increase of the overall process energy consumption, since the additional heat load of the column is neutralized by heat recovery. Assuming the reflux ratio of column in Figure 73 is further increased, a new pinch is generated at the temperature of the distillation evaporator. If this occurs, further reflux increases will result in higher energy requirement of the overall process as well.

This insight can be directly applied to heat pumps. Since it is detrimental to transfer heat from above to below the pinch temperature, it should be beneficial to transfer heat from below to areas above the pinch. The rule becomes clear, when the heat/sink source logic is applied to the problem. In Figure 74, the use of a heat pump to transfer heat from process areas below to above the pinch is demonstrated. The heat transferred by the heat pump reduces heat and cooling requirement of the process. Of course, power is necessary to operate the heat pump.

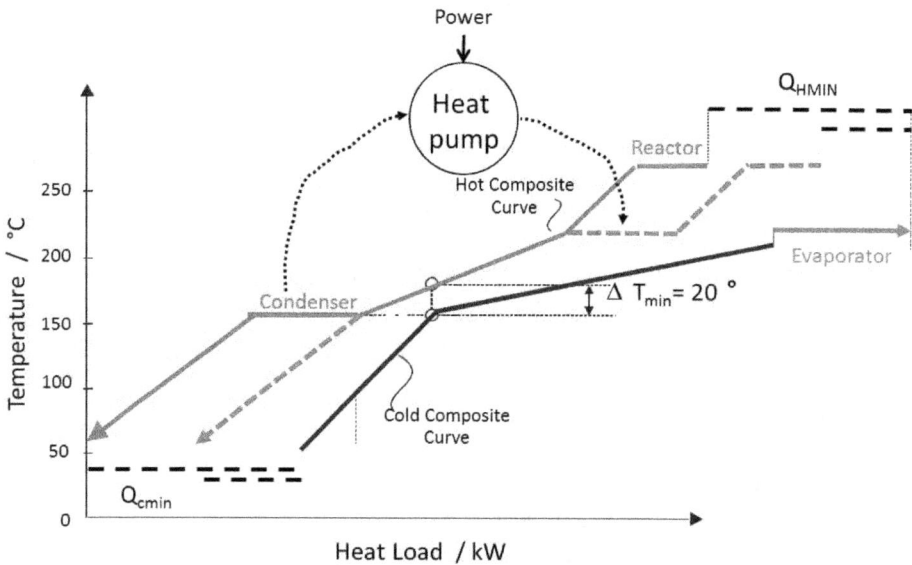

Figure 74: Integration of Heat Pumps

A heat pump upgrades surplus heat below the pinch temperature to temperature levels above the pinch with a heat deficit. Figure 74 shows the impact in a pinch diagram. The heat upgraded by the heat pump (800 kW) reduces the heating requirement. The pinch temperature is still 170°C. Provided the heat pump upgrades the heat only to 180°C, this establishes a new pinch temperature at 190°C. Now there is still no heat transferred above the pinch resulting in an even worse situation – same energy requirement as before, but additional power cost for the heat pump.

Similar considerations govern the integration of heat machines (such as gas turbines) to convert heat into power. These devices have to be integrated into a chemical process either above or below the pinch temperature.

The integration of distillation reactors, columns, heat pumps, or heat machines into a process design can sometimes shift the pinch temperature resulting in changing heat sinks and sources. It is necessary to consider this when designing a process.

The pinch point methodology provides additional insights into the process. Energy-consuming distillation columns and endothermic/exothermic reactors, heat pumps and power generators are operations that can affect the energetic performance of a chemical plant significantly. The pinch methodology gives valuable advice how to use and position this energy relevant equipment into chemical plants.

Capital Targets

Obviously, heat can only be transferred as long as the temperature of hot stream section is higher than the cold stream section. The area where the temperature difference is minimal is called the pinch. Actually, the pinch temperature defines the minimal temperature difference feasible for heat recovery. The smaller the pinch temperature difference gets the more heat can be recovered.

The complete heat transfer area for a process can be estimated according to the formula given in Figure 66.

The composite curve provides a target for the heat and cooling requirements and the heat transfer area as well. The area can be used to estimate the cost of the heat recovery systems without actually designing a heat exchanger network. Both targets depend on the pinch temperature selected (Figure 75). A pinch temperature difference of 0°C requires an infinitely large heat exchanger area at the absolute minimum utility consumption. [10]

The larger the temperature difference is chosen, the smaller the area and cost of heat exchangers become, while the heat recovered decreases as well resulting in higher heating requirements (and cooling demand). Figure 75 illustrates a pinch point diagram including three different pinch temperatures (10/25/40°C). Increasing pinch temperatures result in increased requirements for heating and cooling and reduced capital investment.

If the pinch temperature is reduced, the energy consumption (less energy cost) can be minimized, while heat recovery area (more investment) increases – a classical optimization challenge. Figure 76 schematically describes capital investment and utility consumption as a function of the pinch temperature.

This optimization, however, can be quantified before the heat recovery network is designed for the process. In other words, the pinch point methodology provides targets for utility consumption and heat exchanger network areas in advance that

---

[10] *Composite curves can be shifted horizontally but not vertically.*

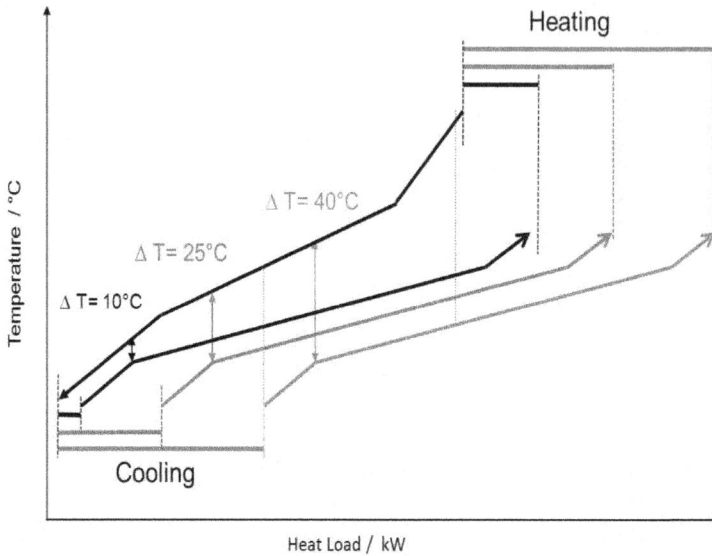

Figure 75: Energy and Capital Targets

later are used to guide the process design. Key for the targets becomes the selection of the optimal, minimum pinch temperature (Figure 76).

The pinch point methodology provides two different targets at the optimal pinch temperature:

- *An energy target describes the minimum energy necessary to drive the chemical process (including the cooling requirements).*

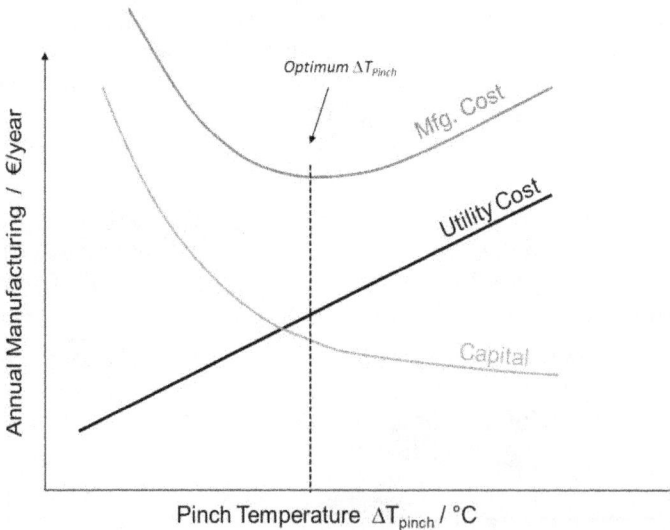

Figure 76: Trade-off between Capital Investment and Energy Consumption

- *A target for the heat exchange area defines the minimum exchange areas.*

The technical targets are translates into economic potentials applying energy and cooling utility prices and heat exchanger and equipment cost.

In addition, the pinch point methodology provides guidelines how to integrate reactors, distillation columns, and heat pumps into the process.

Furthermore, these recommendations are helpful to design the heat exchanger network to recover heat within a process (described later in chapter 3.4.2).

The pinch point methodology can also be translated into heuristic rules summarized in chapter 5 (level 5a).

Material Utilization Target

The material utilization target depends on the synthesis route selected to convert raw materials into a desired product.

Different methods have been proposed to derive from the formula of the desired product the optimal synthesis route. These methods are not part of this book, since process designers normally start with one or sometimes a few synthesis routes proposed by chemical development.[11]

Provided a synthesis pathway was chosen, the material utilization target is simply given by the stoichiometric conversion of the raw materials into a product assuming no losses during product synthesis and recovery.

Target Process Yield (%) = 100% (based on stoichiometric balance)

Actual Process Yield (%) = Actual Reaction Yield * Actual Processing Yield[12]

Target Material Use = kg Raw Material / kg Product (based on stoichiometric balance)

The target for the overall process yield is given by the stoichiometric balance, while experimental data are required to calculate the actual material usage leading to the actual yield. Generally, the actual process yield is the product of reaction yields, recovery yields and formulation yields according to the basic structure of a chemical plant. Target process yield assumes a stoichiometric reaction yield and no losses during recovery and formulation.

Experimental data from laboratory or pilot studies or the actual process provide the actual process yield. The technical data can also be combined with cost data to calculate the economic potential with respect to raw material utilization.

It is worthwhile to emphasize that targeting and benchmarking is definitely not rocket science. Nevertheless, it is surprising that these concepts are seldom applied in a strict and comprehensive matter. In many cases, the targeting and

---

[11] Reaction pathway selection actually is _creative_ effort like process synthesis.

[12] Modified descriptions are possible.

benchmarking aspect is neglected during process design. Engineers often move too early from conceptual design to detail engineering.

The simple procedure to calculate targets and economic potential is illustrated using the information on reaction conversion, selectivity, and recovery yields provided by the process developers:

Reaction Data:

$A \rightarrow R \rightarrow S$ R desired Product

Rate: $\quad r_A = -k1\ e^{-E1/RT}\ cA$

Rate: $\quad r_R = k1\ e^{-E1/RT}\ cA - k2\ e^{-E2/RT}\ cR$

Rate: $\quad r_S = k2\ e^{-E2/RT}\ cR \quad E1{>}E2,\ k_2/k_1{=}0.1$

Lab Conversion: 75 %

Lab Selectivity: 80%

Separation Data:

Boiling points (at ambient air pressure)

$T_A{=}100°C \qquad T_R{=}130°C \qquad T_S{=}105°C$

Recovery yield R: 95%

Recovery yield A: 100%

Cost Data: Cost A = 40 €/kg, R = 100 €/kg, S no value

Plant Capacity: 120 tons/year of Specialty R

A plug-flow reactor is the reactor of choice based on either use heuristics of chapter 5 or a look in your chemical engineering book (Ovenspiel, 1972). A quantitative analysis leads to a selectivity of more than 90% at a conversion of 75% for a ratio of k_2/k_1 of 0.1 (Ovenspiel, 1972). The actual selectivity of only 80% indicates that there is a potential for improvement.

An analysis also shows that selectivity decreases with conversion rate. Smaller conversion rates lead to better yields. Reactor size becomes smaller resulting in less capital investment. Unfortunately, separation and recycle effort will simultaneously increase. Selectivity, however, decreases only very slightly up to a conversion of 75% for $k_2/k_1{=}0.1$ confirming the 75% conversion as design point for the plug flow reactor.

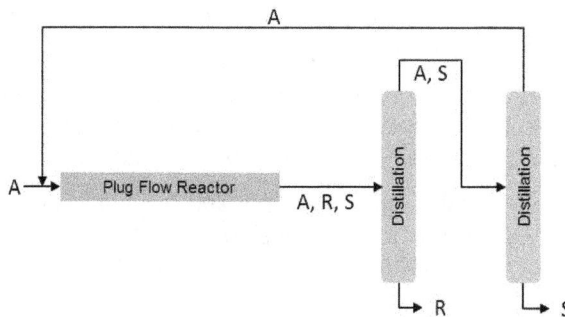

Figure 77: Process Design $A \rightarrow R \rightarrow S$

The plug-flow reactor is followed by a 2-step distillation sequence described in Figure 77, assuming separation of A and S is more difficult and best-done last in a smaller column. The recovery yield theoretically could be above 99% for A and R, provided the design of the distillation columns is appropriate.

Improving recovery yield gives smaller raw material losses, but higher capital for separation equipment and more energy consumption. The designer's choice of at least 95% recovery yield sounds realistic as an initial choice.

The process yield target become

Reaction Selectivity:	90%
Conversion:	100% due to recycle of A
→ Reaction Yield:	90%
Recovery Yield:	99%
→ Process Yield:	89%

The actual process yield is only

Actual Recovery Yield:	80%
Actual Recovery Yield:	95%
→ Actual Process Yield:	76%

The actual raw material usage is about 158 tons/a compared to a target usage of only 135 tons/a feedstock A to produce 120 tons/a product R.

The economic target of this yield gap sums up to

$$\Delta ET \text{ (Materials)} = \text{(Actual Usage – Target Usage)} * \text{Price/Unit} = 0.93 \text{ Mio. €/a}$$

This economic target on material usage provides the financial potential for process optimization or capital investment to close the performance gap.

Economic Potentials

The economic potential of a process design can be calculated adding up the different economic targets. Economic targets and overall economic potential can be linked to the levels of a hierarchical design methodology and guide process design. Various concepts to apply economic potentials for process design have been proposed in the literature (Douglas, 1988), (El-Halwagi, 2012). Douglas proposed a multi-level decision strategy for process designs based on economic potential (Douglas, 1988). [13]

In Figure 78 a five level approach to process design is described modifying the concepts of (Douglas, 1988) and (El-Halwagi, 2012). The economic evaluation becomes more detailed with each new level. Each level adds new design elements to the process design.

Level 1 only considers feed, waste and product cost/prices to calculate an economic potential EP1. This potential EP1 is calculated for the actual design and

---

[13] *The definition of economic potentials always follows the specific design task.*

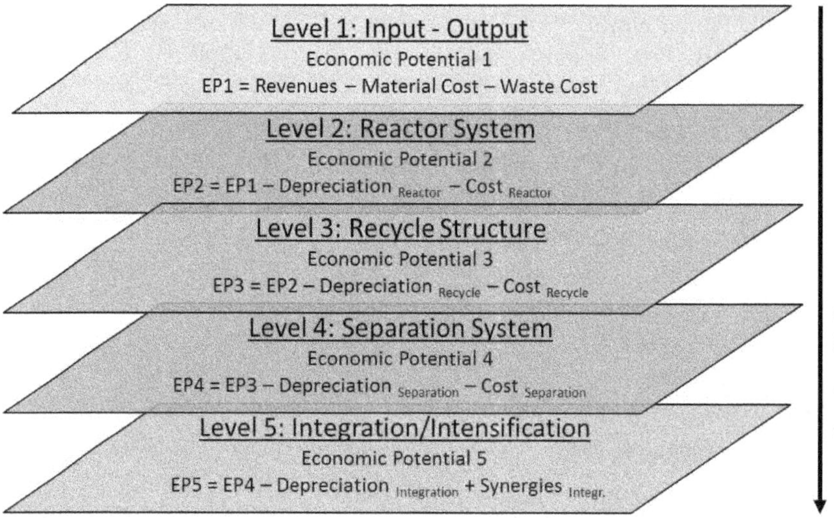

Figure 78: Economic Potential Approach for Process Design

compared with the economic target of material usage. EP1 is also used to study the influence of raw material cost and product price as well.

The next level EP2 takes into account depreciation and operational expenses for the reactor system. EP2 characterizes the economic potential when material cost and reaction related cost are subtracted from product revenues.

Level 3 includes the recycle expenses to EP2 resulting in an EP3. Level 4 adds separation cost to generate an economic potential EP4. The initial process design is reflected by the economic potential EP4. EP4 is comparable to a gross margin (= revenues − manufacturing cost). At level 5, the economic potential introduces investment depreciation and savings due to process intensification and integration measures.

Technical and economic targets are considered at all five levels to rate alternative designs. Economic potentials describe design profitability at each level. Generally, an economic potential can be specified for every level. All economic potentials need to be positive with EP1>EP2>EP3>EP4. Level 5 adds efficiency to the design improving EP5. Projects on level 5 are only pursued when EP increases.

Design Case 8 − Economic Potentials

The procedure of a hierarchical design approach using economic potentials is demonstrated using the earlier example A → R → S.

Economic potential EP 1:   Raw material cost A: 6.32 Mio. € / year
                                    Product revenues R: 12.0 Mio. € / year
                                      → *EP1 = 5.68 Mio. € / year*

Economic Potential EP2:	Reactor Investment:	2 Mio. €
	Depreciation:	0.2 Mio. €/a (10 years)
	Operation reactor:	1.5 Mio. €/a

→ *EP2 = 3.98 Mio. €/a*

Economic Potential EP3:	Recycle Investment:	1 Mio. €
	Depreciation:	0.1 Mio. €/a (10 years)
	Operation recycle:	0.5 Mio. €/a

→ *EP3 = 3.38 Mio. €/a*

Economic Potential EP4:	Separation Investment:	5 Mio. €
	Depreciation:	0.5 Mio. €/a (10 years)
	Operation separation:	2 Mio. €/a

→ *EP4 = 1.88 Mio. €/a*

Economic Potential EP5:	Integr./Intens. Invest:	2 Mio. €
	Depreciation:	0.2 Mio. €/a (10 years)
	Operation savings:	1.32 Mio. €/a

→ *EP5 = 3 Mio. €/a*

The process design in Figure 77 provides an economic potential EP5 of 3 Mio. /a. corresponding to a return on sales of 25% and a cash flow of 4 Mio. €/a for the project. These numbers do not include any expenses for marketing, R&D or overhead & allocations, but already indicate a profitable business.[14]

The economic target gap ΔET for raw materials amounted to 0.93 Mio. €/a supplementing the economic potentials 1-5.

Technical targets, economic targets, and economic potentials guide the process designer searching for the optimal process design.

Obviously, the concept of targets and potentials has to be applied to process design in a smart way – flexible and adjusted to the individual design task.

---

[14] *A more detailed evaluation gives chapter 4.2.*

3.3 Technology Opportunities – It's about Process Intensification

Chemical processes have continuously been improved during the last years. Unit operations have reached a high level of sophistication. Future improvements are not likely to result only from a better design of unit operations and equipment, but require a synergetic approach. Vertical synergies across macro-meta-micro-nano-scales and horizontally integrated chains of unit operations offer opportunities for better efficiencies.

Process Intensification across effect levels addresses the vertical dimension. *Process Integration* covers the horizontal dimension. In other words, process intensification mainly occurs on a device level, while process integration deals with complete process systems – up to total supply chains. Figure 79 illustrates the system approach of intensification and integration of processes.

Intensification of chemical processes aims at a significant, economic, and ecological efficiency increase of chemical and biological processes and the creation of novel or improved product qualities. Integration of chemical process

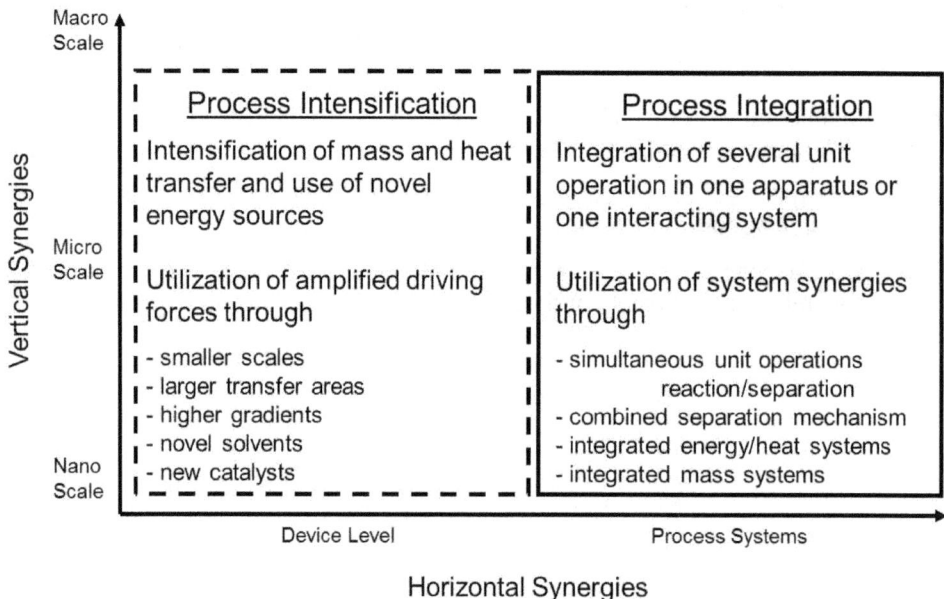

Figure 79: Process Intensification and Integration

aims at better process efficiencies benefiting from better-integrated processes

Process intensification and integration represents not a new technical discipline, but an extremely focused methodology approach. Efficiency with respect to material utilization, energy consumption, and waste avoidance directly links process intensification and integration with the drive for industrial sustainability.

3.3.1   Intensified Heat and Mass Transfer

Transfer, conversion, and/or generation of mass and energy form the core of process intensification. Mass transfer during mixing is essential to establish the proper reactant ratios for high yields. Removal of reaction heat is similarly important for selectivity/productivity and safety as well. An efficient heat and mass transfer is particularly important for highly exothermic and fast reactions.

Figure 80 gives an overview on different reaction classes characterized by heat of reaction and reaction half time. Nitration and hydrogenation are commercially important reaction categories in the chemical industry. Metal-organic Reactions and decompositions are examples for strongly exothermic reactions.

Figure 81 describes the heat and mass transfer topic for a stirred reactor requiring heat transfer into the reactor and mass transfer from a continuous into a dispersed

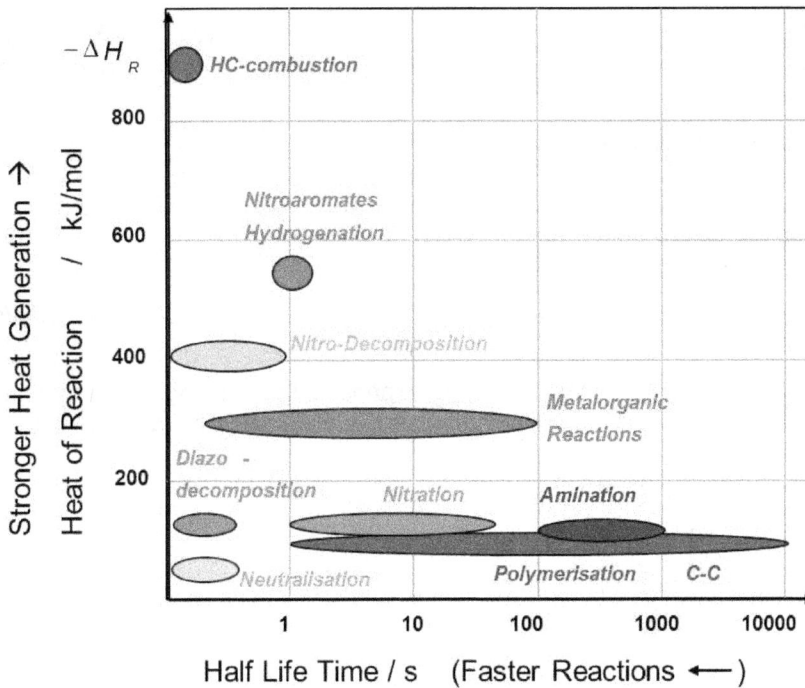

Figure 80: Overview on Reaction Classes

phase. Process intensification aims at maximizing heat and mass transfers defined by the driving forces, transfer areas, and transfer coefficients of heat or mass.

Process intensification looks for technical solutions that provide particularly smart configurations to accomplish that task. The favorite topic for process intensification has been heat transfer for many years.

Heat Transfer

The heat transfer rate Q (J/s or W) increases with larger temperature difference, exchange area and heat transfer coefficients. Process designers try to enhance these drivers.

$$Q = k\, A\, \Delta T \qquad \text{with } \frac{1}{k} = \frac{1}{\alpha a} + \frac{\delta}{\lambda} + \frac{1}{\alpha i}$$

Turbulent flows improve the heat transfer coefficient compared to laminar flows. Geometrical designs multiply the equipment exchange area. Counter-current design maximizes the driving force temperature. Counter current arrangements are consequently a preferred solution in any process-intensified concept.

Particularly heat exchange area, flow characteristics and counter-current features distinguish alternate processing equipment such as stirred reactors, multi-tube reactors, or micro-reactors.

Intensified heat exchange generally reduces the time necessary to remove or add heat to a reaction volume. The material is significantly less time exposed to high temperatures.

Figure 82 qualitatively illustrates the relation between heat transfer rates, residence time and heat exchange area for different reactors. A micro-reactor has 1000 times more exchange area than a stirred tank dramatically increasing heat transfer rates at a comparable temperature driving force. Residence time to heat a certain volume drops from hours to milliseconds.

Process intensification results in less product damage, smaller equipment, and/or safer operations.

Since process intensified equipment is more expensive than standard equipment, a switch to intensified equipment makes only sense, if an operational, economic or

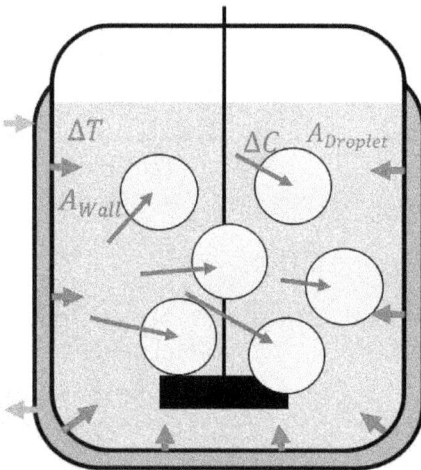

- Heat Transfer Q (J/s or W)
  - $Q = k_H\, A_{wall}\, \Delta T$

- Mass Transfer M (kg/s)
  - $M = k_M A_{droplet} \Delta C$
  (from continuous phase into droplets)

- Increase of
  - Area $A_{droplet}$, A_{Wall}
  - Driving Force ΔC, ΔT
  - Transfer Coefficient k_M, k_H

  improves transfer rates

Figure 81: Heat and Mass Transfer in Stirred Reactors

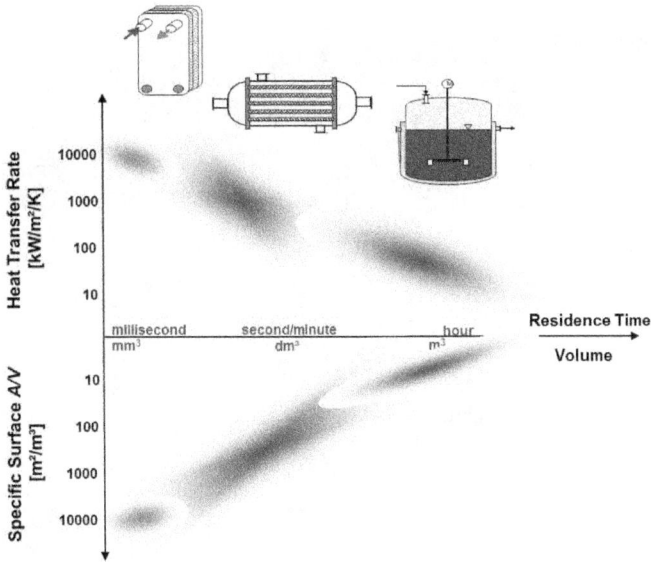

Figure 82: Comparison of Heat Transfer Equipment

ecological benefit is expected.

In fine chemicals, micro-reactors increasingly perform heating, cooling and fast, exothermic reactions due to their highly efficient heat and mass transfer features.

Figure 83 shows a micro-reactor developed by the Lonza currently distributed by Ehrfeld. The micro-reactor consists of multiple reaction and heating / cooling plates resulting in an easily expandable capacity.

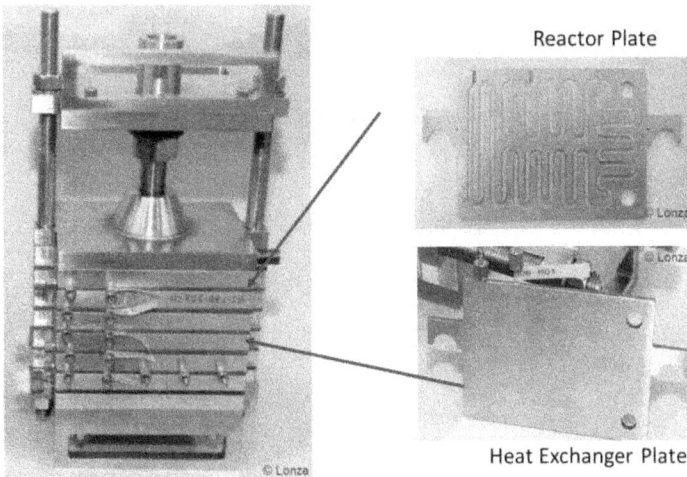

Figure 83: Micro-Reactor (Source: Ehrfeld Mikrotechnik BTS)

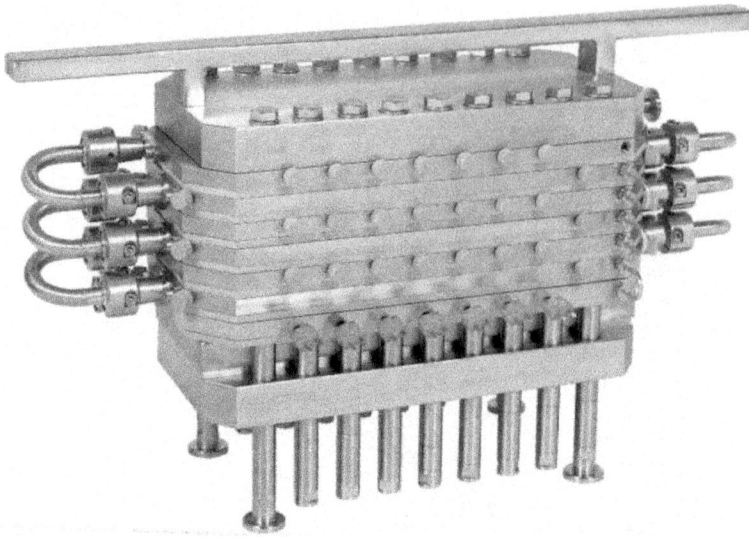

Figure 84: Plate Micro-Reactor (Source: Ehrfeld Mikrotechnik BTS)

Figure 84 shows a micro-reactor design structured similar to heat exchangers. The flow channels are only a few hundred micrometers wide compared to heat exchangers with 1 to 5 mm channel widths. Alfa-Laval developed this design distributed by Ehrfeld as well.[15]

Of course, the micro-reactor is more expensive and requires some maintenance to avoid plugging. The lower space requirement of the compact micro-reactor compared to a batch reactor is beneficial. Productivity and yield improvements add to these benefits (depending on the reaction pursued).

In general, the design is easily scaled up to large-scale applications. A residence time of a few seconds can be realized at large heat transfer rates for flow rates of several tons per hour.

The application of micro-reactors for liquid heating and evaporation is not limited to small flow rates. Particularly heating of liquids and gases are regularly performed with millimeter / micrometer structured devices for various processes.

Figure 85 shows a pilot reactor of Bayer consisting of multiple reaction channels. The modular design allows an easy scale up to several tons/hour. This is an advantage of scaling up by numbering up characteristic for modular designs.[16]

Specific inserts in micro-structured channels enable large heat exchange areas, excellent heat transfer coefficients, small hold-ups, and short residence times.

---

[15] www.ehrfeld.com

[16] Heck. (2012). Verfahrenstechnik, pp. 1-2.

Miprowa Technology

Figure 85: Heat Exchanger with Micro-Structures (Source: Ehrfeld Mikrotechnik BTS)

These feature result in superior product quality due to less product residence time at high temperature.

BASF uses a micro-structured heat exchanger manufactured by heatric transferring heat of more than 1 MW in the EOX process. This heat exchanger possesses a heat transfer capacity of 1MW: Once more, a huge heat transfer rate can be accomplished at a small hold-up leading to short residence-times.[17]

DSM together with KIT has developed a reactor with a heat transfer capacity of

Figure 86: Ceramic Micro-Reactor (Source: Corning)

[17] www.heatric.com, http://www.dpxfinechemicals.com, www.corning.com,

100kW for fine chemical production. The hydraulic capacity is up to 2 ton/h for a liquid feed. The advantages are huge heat exchange areas, small residence times and as an option higher temperature of the heating medium due to lower product damage. High exchange rates are the result.

Finally, Corning offers modular micro-reactor systems made of glass and ceramic materials (Figure 86).

The application of process intensification depends on five questions:

- Does the operation benefit from better heat transfers at all?
- How to increase heat transfer areas? Are there any consequences on the process?
- How to influence heat transfer coefficient positively? What is the transfer rate-limiting step?
- Is the temperature difference optimally used? Is a counter current mode preferable?
- Is the larger investment in equipment justified by higher returns?

Mass Transfer

The mass transfer topic for process intensification is relevant to transfer of a component from one medium into another or separate a mixture into their components. Transferring a component from one medium into another is comparable with the heat transfer problem.

$$M = k \, A \, \Delta C \qquad \text{with} \quad \frac{1}{k} = \frac{1}{\propto a} + \frac{\delta}{\lambda} + \frac{1}{\alpha i}$$

Temperature difference is replaced by concentration gradient and heat by mass transfer coefficient. The interface area stays in the game.

Extraction, absorption, adsorption, chromatography, membrane techniques, and distillation are the most popular separation technologies. Separation technologies particularly benefit from counter-current concepts provided driving force need to be enhanced.

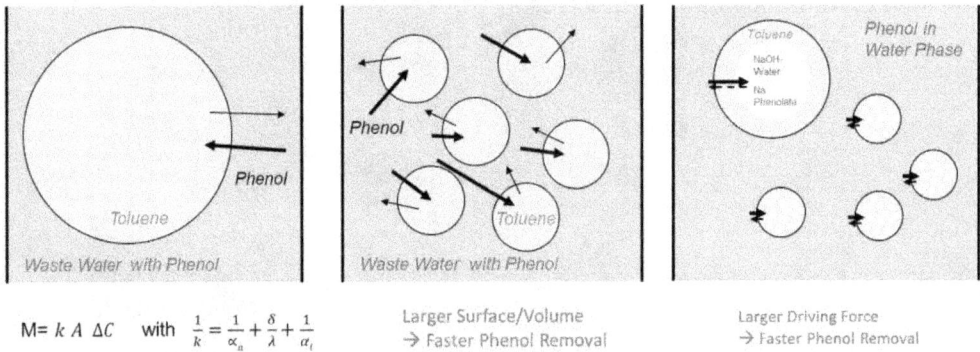

$$M = k \, A \, \Delta C \qquad \text{with} \quad \frac{1}{k} = \frac{1}{\propto_a} + \frac{\delta}{\lambda} + \frac{1}{\alpha_i}$$

Larger Surface/Volume
→ Faster Phenol Removal

Larger Driving Force
→ Faster Phenol Removal

Figure 87: Extraction with smaller Droplets

Extraction is a unit operation that is governed by mass transfer rules. In a 2-phase system there are droplets dispersed in a continuous phase. Normally a component included in one phase will be extracted into the other phase.

A standard example for such an extraction process could be the removal of phenol from wastewater using toluene as receiving phase. The equilibrium distribution of phenol between toluene and water is in favor of toluene. The mass transfer rate of phenol from water to toluene can be increased by larger interface areas, higher concentration gradients, and improved mass transfer coefficients.

Consequently, a decrease in droplet size could significantly increase the mass transfer area while still using the same volume ratio of continuous and dispersed phase (Figure 87 – middle scheme). A reduction of the toluene droplet diameter by a factor of 100 from 1 mm to 10 micrometer will increase the droplet surface area and the transfer rate significantly provided other parameters are unchanged.

How can we enhance the driving force? The concentration gradient of phenol is determined by the difference between the concentrations of that component in both phases. This gradient affects the process efficiency in two ways. First, the concentrations in both phases determine the mass transfer rate gradient until the equilibrium is achieved. Secondly, the equilibrium defines the final distribution of phenol in both phases.

Reactive extraction with a reactive component dissolved in toluene that has an affinity to phenol could improve the equilibrium. In Figure 87 (right side), an extraction concept is outlined that uses an internal water phase in the toluene

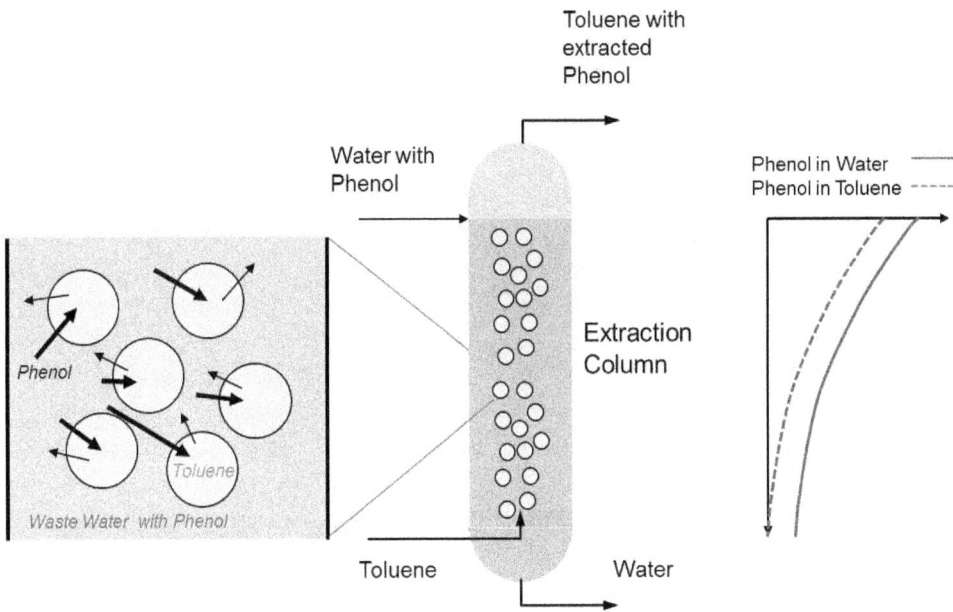

Figure 88: Counter-current Operation

droplet. This inner water phase contains sodium hydroxide as reactive ingredient immediately converting phenol into sodium salt of phenol completely.[18] The phenol concentration in the inner phase, therefore, is zero generating the maximal driving gradient during the extraction process. A reactive component that forms a complex with phenol represents an alternative to sodium hydroxide.

Finally, the driving concentration Δc_{Phenol} can be increased by applying a counter-current flow in an extraction column. This is schematically described in Figure 88. A counter-current operation enables a maximal driving force across the column length. In particular, the water at the bottom of the column loaded with little phenol is extracted with fresh toluene leading to the lowest phenol concentration possible.

Finally, the diffusion process at the droplet interface influencing the mass transfer coefficient can be enhanced by mixing or internals inside the extraction equipment. Diffusion within the toluene phase could be enhanced by shorter distances between the two water phases or higher temperatures increasing diffusivity.

In general, the removal of the dispersed toluene droplets from the continuous water phase by gravity forces becomes more difficult for smaller droplets. Therefore, different or better separation techniques are necessary.

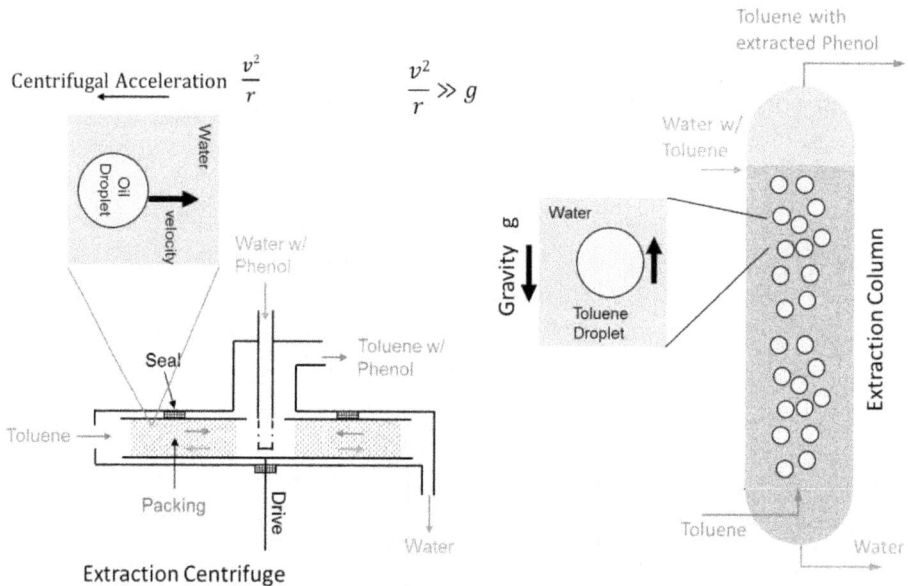

Figure 89: Centrifugal versus gravitational Operation

---

[18] *The sodium hydroxide containing toluene droplets have to be stabile during extraction. This represents a challenge to emulsion breaking later on.*

Feed (3 Phases s/l/l)

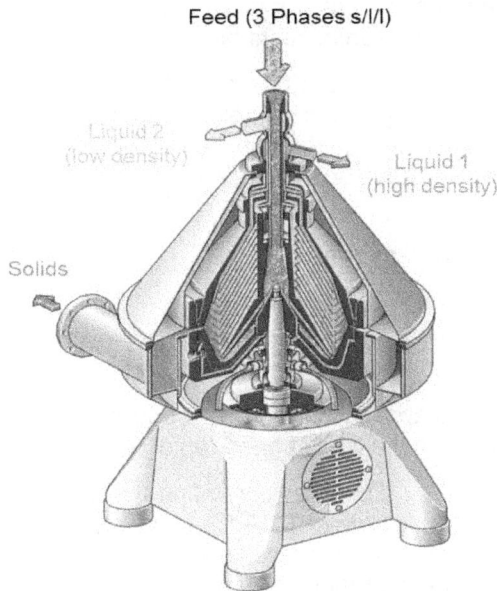

Figure 90: Extraction Centrifuge (Source: GEA Westfalia Separator Group GmbH)

Process intensification for phenol extraction requires a stronger gravity field during extraction. The counter current extraction process normally performed in large extraction columns are not applicable to very small toluene droplet systems. These can be reached in the centrifugal field of hydro-cyclones or centrifuges. Gravity in centrifuges can exceed the natural gravity by a factor of up to 100.

The relative velocity difference describes the separating velocity between continuous phase (water) and droplets (toluene):

$$v_{rel,\ water-toluene} \sim m_{droplet}\ v_{rotation}{}^2\ /\ r >> m_{droplet}\ g$$

The same relative velocity is reached for smaller droplets in a centrifugal field compared to a gravity field. Extraction centrifuges are the process intensified alternative of extraction columns.

Figure 89 compares the extraction process in a continuously operated column with a continuously operated centrifuge. In an extraction column, separation is driven by gravity. In a centrifuge, the separation of both phases is enhanced by a high gravity field.

Figure 90 shows the set-up of a centrifuge used for extraction. The advantage of centrifuges are better mass transfer and droplet separation, less space requirement and less sensitivity to fouling compared to columns. More maintenance (moving parts), complex machine, and higher investment present the disadvantage.

In addition to significant larger mass transfer rates, centrifuges possess little hold-up. Start-up and control of such units might be easier as well.

The phenol extraction also confirms a property of process intensification. Process intensification normally leads towards complex systems. The phenol extraction in the process-intensified version requires a more efficient separation process for the toluene droplets and an additional breaking process for the water droplet containing toluene phase.

Design Case 9 – DOW Process

Fast, multiphase reactions require efficient mass and heat transfer, mixing, cooling, or heating capabilities. In general, the same principles that are applied to separation are beneficial to improve reaction performance as well.

DOW realized an industrial application for the production of hypochlorous acid. The product HOCl is not stable in the liquid phase rapidly decomposing into chlorate.

The solution developed by Dow is a rotating packed bed permitting rapid absorption, reaction, and desorption of HOCl. In a high g centrifugal field, gas-liquid mass transfer is strongly enhanced. This process-intensified design reduces the size of the processing equipment by a factor of 40. Yield increases from 70 to more than 90%. Stripping gas is reduced by 50%; wastewater and chlorinated byproducts are reduced by 30%. The environmental impact is significantly smaller – a sustainable solution.

A scheme of the reactor set-up is shown in Figure 91. The equipment size of the conventional and the centrifuge-based approach is compared in Figure 92 confirming the benefits in terms of space, hold-up (safety), and investment.

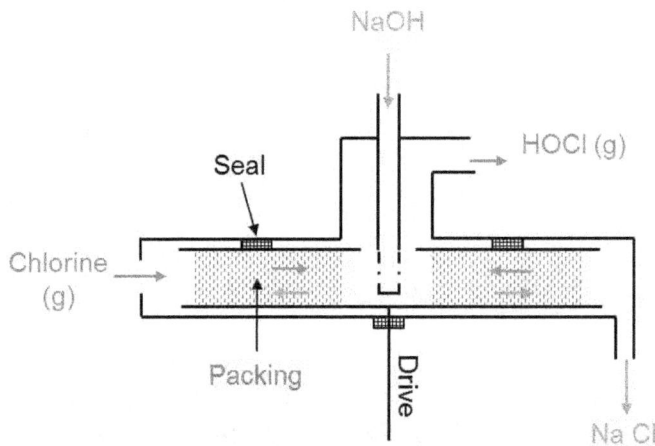

Figure 91: Scheme of Centrifuge Reactor (gas-liquid)

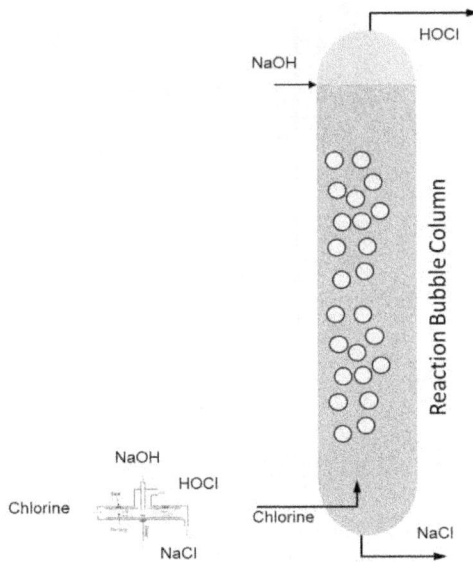

Figure 92: Comparison of HOCl Process Equipment Equipment

Design Case 10 – Intensified Fermentation

Oxygen supply for fermentation is classical mass transfer problem in chemical engineering. Conventionally oxygen is provided to the fermentation broth using air. Oxygen supply from air bubbles into the fermentation broth is accomplished through air sparking devices generating small bubbles rising in the fermenter. Once more smaller droplets will provide more transfer area resulting in higher mass transfer rates. In addition to area increase, a switch from air to pure oxygen will increase the partial oxygen pressure leading to a better driving force (Figure 93).

In fermentation, the supply of oxygen is accompanied by a stripping of carbon dioxide. Both fluxes will be amplified with smaller droplets and pure oxygen. Oxygen consumption and carbon dioxide production are determined by the metabolic activity of the microorganism.

Batch operation represents the standard procedure for fermentation facilitating a sterile operation. Semi-batch or continuous operation can provide opportunities for process intensification. A further improvement of productivity can be achieved when biomass is recycled. Of course, the specific fermentation features determine whether continuous operation with biomass recycle is feasible.

Figure 94 demonstrates impressively the benefit of a process intensified fermentation. Continuous fermentation combined with biomass recycle increases the space yield by a factor of 16, the residence time of the product protein in the fermenter is reduced by a factor of 4.

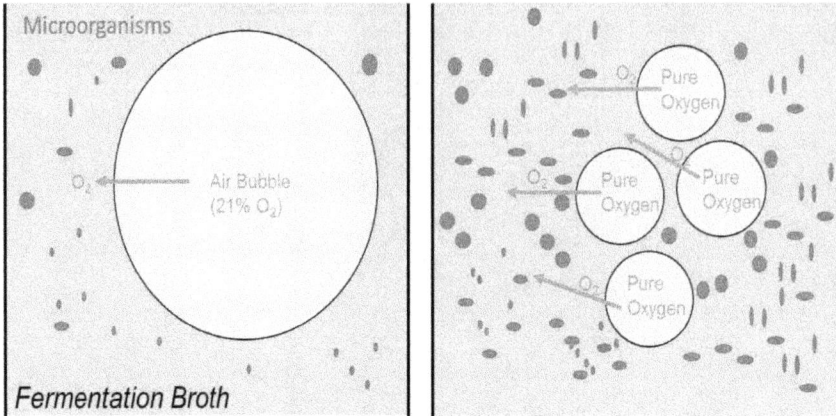

Figure 93: Oxygen Supply in Fermentation

Continuous operation with biomass recycle, however, becomes mandatory, if the product is not stable in the fermentation broth. Under these circumstances product residence time in the fermenter has to be minimized using continuous operation. Biomass recycles is mandatory to achieve an economically necessary productivity.

Microorganisms used for fermentation are sensitive, biological systems that are affected by mechanical stress. This limits the use of machines to generate additional interface for mass transfer. Chemical systems, however, do not suffer from that constraint.

Once more, the proper questions concerning intensification of fermentation are:

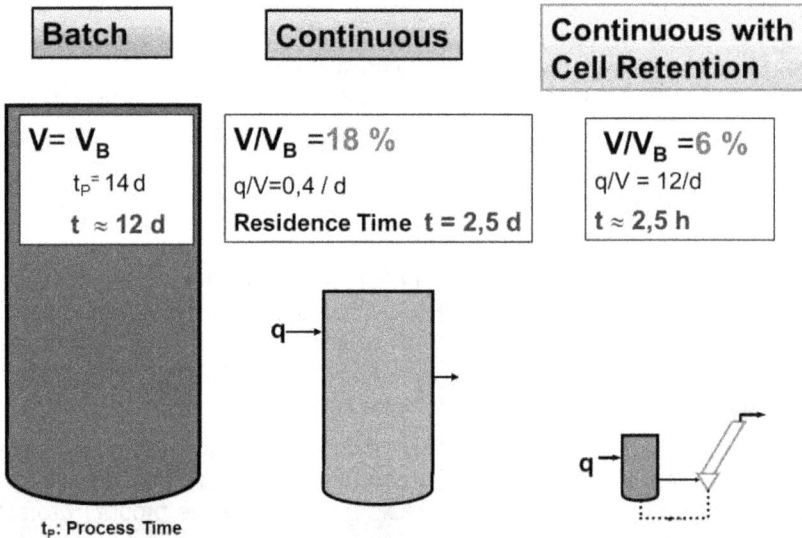

Figure 94: Comparison of Fermentation Operation Modes

- Does the operation benefit from an appropriate oxygen supply and carbon dioxide removal I?
- How to increase oxygen transfer areas?
- How to influence transfer coefficient positively? What is the transfer rate-limiting step?
- What is about the concentration gradient (pure oxygen)?
- Are there any consequences on the process (mixing still sufficient, carbon dioxide removal, cell survival)?
- Is there an economic advantage linked to intensification of mass transfer?

Distillation in complex Machines

A similar approach is applicable to distillation. Distillation is generally performed in columns with internal structures supporting mass transfer. The development of a structured packing could already be considered process intensification. Packing, however, represents a gradual development. Process intensification aims at a qualitative, disruptive improvement. Distillation performed similar to extraction in a high gravity field could allow a dramatic improvement provided gas bubble size could be reduced significantly.

Increased mass transfer areas through smaller gas bubbles can be used in a counter-current operation when distillation occurs in the gravity field of a centrifuge. Internal structures can also be used in centrifuges to amplify mass transfer rates. In Figure 95, the basic design of high gravity distillation centrifuge is outlined. The design allows packing structures to be modified similar to conventional columns.

Distillation in the centrifugal field of a centrifuge allows process intensification with respect to mass transfer rates due to small bubbles in turbulent flows. Relative velocities between gas bubbles and continuous phase, however, are still linked in

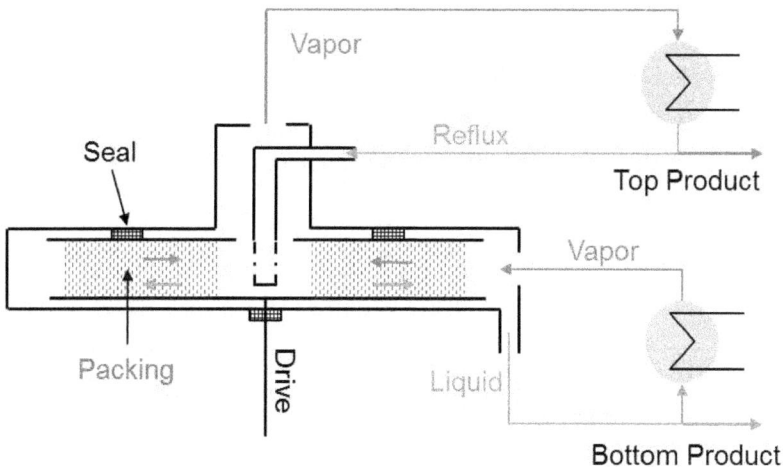

Figure 95: Distillation in a centrifugal field

centrifugal distillation limiting throughput due to flooding. In addition, heat input still occurs in an evaporator unit, cooling in a condenser unit. As a result, the gas-liquid flow ratio cannot be varied in a distillation centrifuge. In a process intensified distillation set-up, a design would be desirable that allowed a "perfect" selection of the gas-liquid ratio and a locally distributed heat input.

Although the concept of operation with centrifugal fields has been promoted regularly during the last 30 years, little progress has been made using this approach in real world problems.

Similar to heat transfer mass transfer can be translated into five questions:

- Does mass transfer affect process performance at all?
- How to increase transfer area? Does an increased area affect process otherwise?
- What is the limiting step? Can design features increase mass transfer coefficients?
- Is there an opportunity to influence the driving concentration gradient? Is it possible to affect the final equilibrium?
- Do efficiency gains justify capital investment?

There are wide literature on process intensification available (Stankiewicz, 2004), (Bradshaw, Reay, & Harvey, 2008).

3.3.2 Intensified Mixing and Reaction

An early topic for process intensification has been mixing – particularly in micro-mixer units. Mixing characteristics are particularly important, when reactions are very fast. Up to now, the selection of the best reactor design was only based on the kinetics of the reaction system (Chapter 3.1.2).

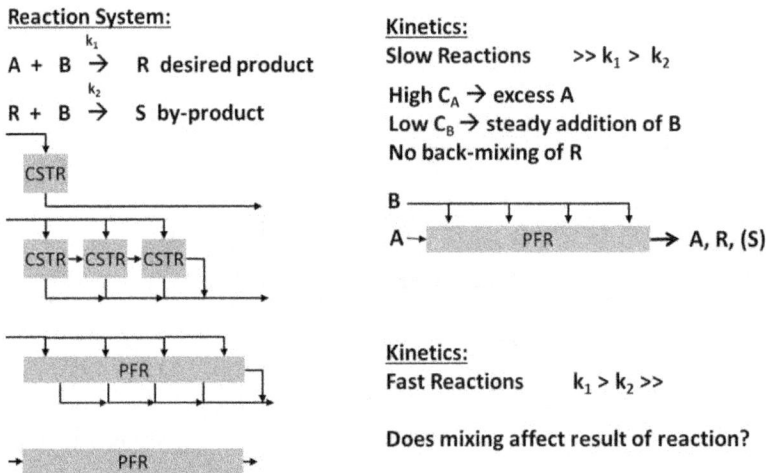

Figure 96: Reactor Design for slow/fast Reactions

Figure 96 describes a reaction system consisting of competing reactions.

Provided the two reactions are slow compared to the mixing time (e.g. $>> k_1 > k_2$), a plug flow reactor is the reactor of choice to minimize the generation of the undesired by-product S. The selectivity towards R is further improved by high concentrations of A and low concentrations of B. This could be accomplished, if an excess of A is used and B is not fed completely at the entrance of the reactor, but steadily along the tube reactor.

If the mixing time in a reactor is in the same range as the reaction half time, reaction kinetics that are always based on perfectly mixed systems are no longer sufficient for the reactor design. What is the approach appropriate for very fast reactions (e.g. $k_1 > k_2 >>$)?

Mixing consists of macro- and micro-mixing. Macro mixing distributes homogenous cluster of molecules of component A in homogeneous clusters of molecules of component B. Macro-mixing is demonstrated with homogeneous layers of A and B in Figure 97. It does not change the conclusion, if vortex clusters of A or B replace these layers.

Macro mixing is driven by flow turbulence and "layering" steadily reducing the "thickness" of these homogeneous layers. In addition to macro mixing, micro mixing simultaneously takes place due to molecular diffusion. Molecular mixing dominates the mixing process as soon as the layers or clusters become thin or small enough. A homogenous mixture of two components is achieved when both components are mixed on a molecular level.

Macro- and micro-mixing is completed for slow reactions, before the reaction quantitatively starts (upper example in Figure 97). The assumption that a homogeneous mixture governs the reaction process is valid and a kinetic approach

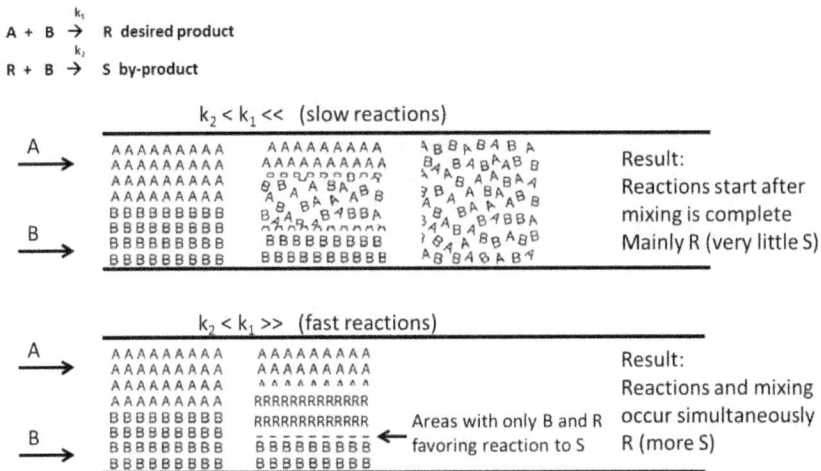

Figure 97: Mixing and Reaction for slow and fast Reactions

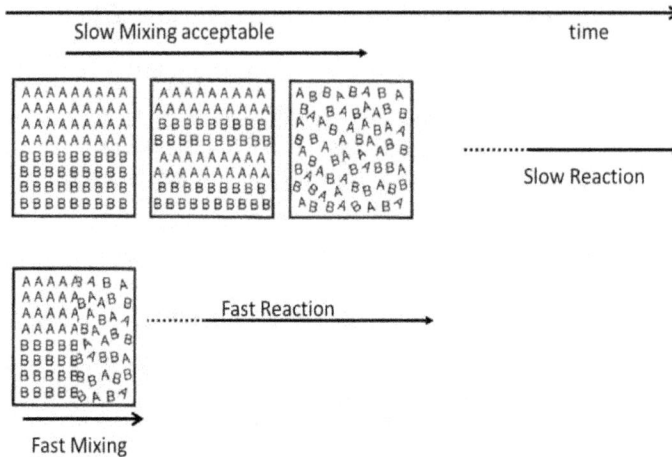

Figure 98: Mixing Feature for slow and fast Reactions

to reactor design is sufficient.

Provided the reactions are fast, mixing and reaction occur simultaneously. Figure 97 (lower part) illustrates that A and B already start reacting at the layer interface between A and B. Significant product quantities of R are generated, while mixing is still on-going. There are still areas with mainly B (or at least little A) and newly generated R due to incomplete mixing of A and B.

In these areas, the undesired reaction of B with R produces S at high rates, since the concentration of component B and product R favor the undesired by-reaction. The assumption that a homogeneous mixture governs the reaction process is not valid due to slow mixing. Fast reactions require fast mixing to allow a reactor design based on perfectly mixed raw materials (Figure 98).

Figure 99 shows the impact of mixing intensity and back mixing on the selectivity using competing dyestuff reactions. The theoretical and experimental data confirm that the selectivity toward R improves with better mixing (e.g. higher mixing intensity) in a jet reactor. A jet reactor creates a back-flow in the areas close to the tube wall due to the sucking effect of the nozzle jet.

This back mixing can be minimized when a properly shaped piece is inserted in the reactor. The selectivity towards R is even further improved. The actual selectivity at high mixing intensity and no back mixing reaches the theoretical best selectivity given by the kinetic data and perfect mixing of the two raw materials A and B.

In a jet mixer, macro mixing is accomplished by turbulence. In micro-reactors, flows generally are laminar, since small dimensions only allow small Reynolds numbers. How can micro-reactors mix feed materials efficiently without turbulence?

Mixing in a cascade micro-mixer will be used to demonstrate macro mixing by a "layering"-approach. In Figure 100, this mixing process is schematically described. There are two components that are sliced into layers that become smaller and

Jet-Reactor:

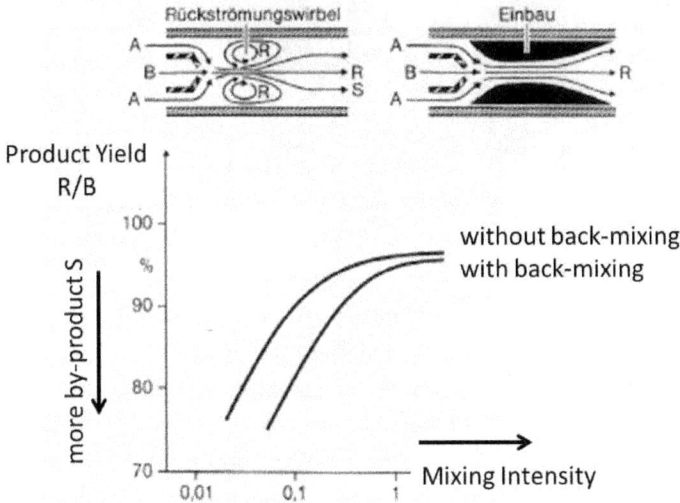

Figure 99: Jet reactors for fast Reactions

smaller. If the process is started with 2 layers of 1 mm each, there are 32 layers of about 30 micrometers after 5 cascades. If we repeated that process indefinitely, we could achieve a perfect mixture (even without diffusion of molecules). In reality, diffusion will "take over" the mixing process as soon as the layers are small enough

Macro mixing is accomplished by cutting big pieces in small slices using mechanical ways instead of turbulence. Figure 100 illustrates a cascade micro-mixers reducing layer thickness in discrete steps into micrometer range where diffusion governs and accomplishes a homogeneous mix. A cascade mixer with five steps reduces the laminar layers from 1000 to below 50 micrometer in milliseconds – without turbulence.

Mixing in micro-reactors represents an interesting example of intensified mass and heat transfer.

Micro-structured devices are particularly suited for cooling/heating, mixing and reaction, if fast, strongly exothermic or endothermic reactions are performed. Hydrogenation, nitration, metal organic reactions, oxidation reactions and many more are well suited for micro-devices.

Organometallic reactions are highly exothermic. Reactions are normally performed in small reactors at small reaction rates at very low temperatures. Removal of huge heat quantities presents a safety challenge, since failure will result in run-away reactions. Micro-reactors offer an excellent heat removal management. As shown earlier this allows reactors with small hold-up. Heat removal and hold-up limit the safety risk dramatically.

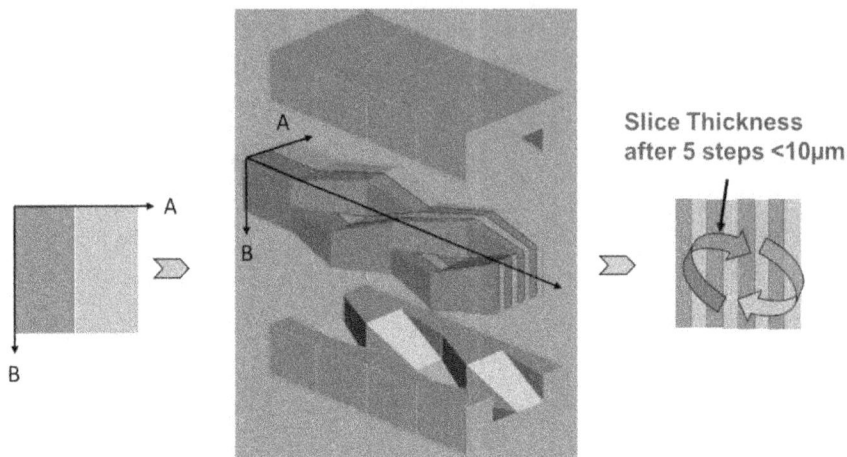

Figure 100: Mixing in Cascade Micro-Mixer

This also allows the reactions to be performed at higher temperatures. In Figure 101, a comparison of an organometallic reaction carried out either in classical stirred reactors or micro-reactors is given confirming the advantages of micro-technology. Safety is significantly improved due to a hold-up of only 2l in a micro-reactor compared to 400l in a traditional tank reactor. The productivity gain results from a 30°C higher reaction temperature (-40° vs. -70°C). Yield is increased from 86 to 94% as a side effect.

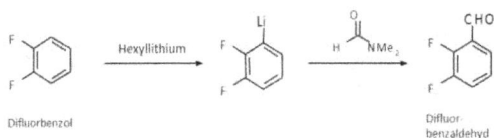

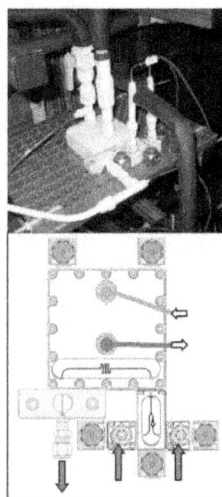

Improved Safety

Batch-Reactor: 400 liters, TR = -70°C

Continuous Micro-Reactor: 2 liters, TR = -40°C

Increased Selectivity:

Batch-Reactor: 86%

Micro-Reactor: 94%

Figure 101: Organometallic Reaction in a Micro-Reactor

3.4 Systems Opportunities - It's about Process Integration

Process integration adds a system perspective to process intensification. Process integration aims at system synergies. These synergies can be generated when heat is recovered and reused in a process or different unit operations are pursued in one apparatus.

The term *process integration* is often used in an inflationary manner. Simply combining two unit operations is not sufficient to justify the term "process integration". Figure 102 shows three systems consisting of reaction and separation. The recycle versions that show a positive influence on the process performance represent rather an optimized process than an integrated process. Process integration of two or more unit operations has to create a unique, new structure with an unambiguously superior performance. A reaction column performing an equilibrium reaction and separation simultaneously represents a truly integrated process version in Figure 102 (center process). A reaction column with a dividing wall combines even two separation columns and one reactor in one device.

Only the system on the right justifies the term process integration, since this

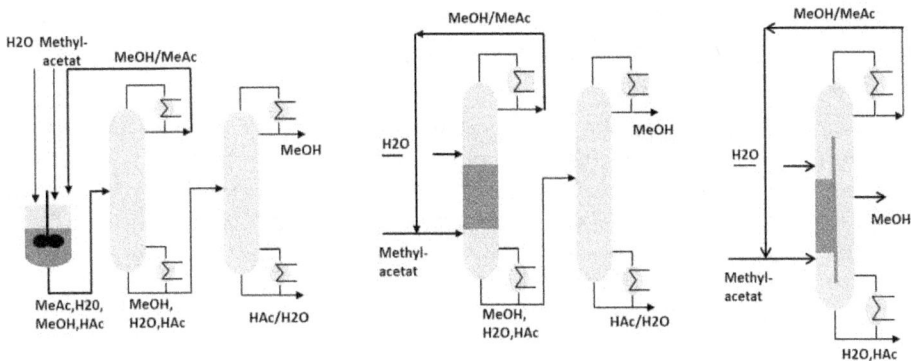

Figure 102: Definition of Process Integration

system approach combines reaction and separation in one device affecting the reaction equilibrium positively. The simultaneously performed reaction and separation processes can significantly improve yield and productivity and result in lower capital investment due to fewer equipment pieces

Process integration includes direct and indirect integration measures. Direct process integration combines unit operations conventionally performed in different equipment units into one device. Indirect process integration links two processing devices through their mass and energy flows.

Figure 103 shows a direct and indirect integrated distillation system. The heat generated in the condenser of one column drives the evaporator of the other column. Of course the temperature levels of both columns have to be appropriate (e.g. $T_1 < T_2$).

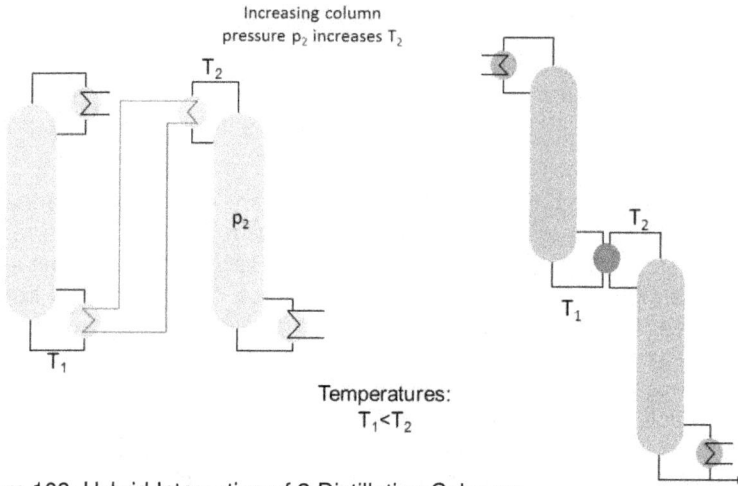

Figure 103: Hybrid Integration of 2 Distillation Columns

The indirect heat integration on the left uses a separate medium to transfer the recovered heat from column 2 to column 1. Direct heat integration combines the heat transfer in one heat exchanger acting as evaporator for column 1 and condenser for column 2.

The advantage of a direct heat integration results from the larger temperature driving force. Heat transfer occurs only once, while an indirect integration uses two units requiring twice a ΔT. This limits the applications for indirect schemes compared to direct designs accompanied by larger capital investment due to two exchangers and additional piping.

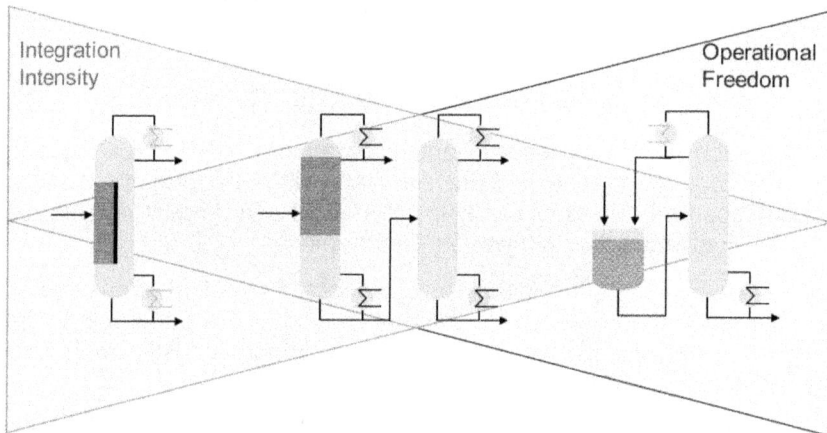

Figure 104: Process Integration, Process Operability and Process Control

Obviously, an indirect integration offer advantages in terms of operability, safety, and special layout.

Truly integrated processes exhibit a more complex dynamic behavior than simply linked processes. The degree of operational freedom is reduced requiring significant process control effort. Deviations from the design parameters can cause severe problems to a reliable operation for an integrated device. The challenges for the process control set-up are significantly larger for an integrated process. Figure 104 schematically outlines this consequence.[19]

3.4.1 Integration of Unit Operations

The interaction between reaction and separation represents the most important area for process interactions. Figure 105 describes a system containing one reactor, two distillation columns, a crystallizer, a solid filter, and a solid dryer. Recycle of feed and solvent couples the different units generating a solid product. A purge and a by-product stream complete this classical flow sheet.

Reactions are very often combined with distillation and extraction for bulk chemicals in a continuous operation. Combinations of distillation, crystallization, and solid filtration are used for fine chemical isolation - operated in batches. If the desired product is a biological molecule such as a protein or an antibody damaged at higher temperatures, membrane technologies and chromatography collaborate with fermentations.

These indirectly linked systems, however, do not represent truly integrated

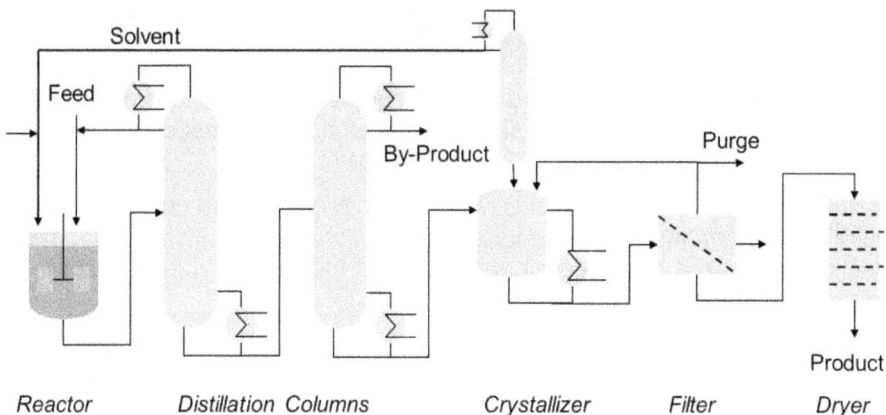

Figure 105: Reaction – Separation – Interactions

---

[19] *Figure 104 is modified version of Agar`s description* (Sundermann, Kienle, & Seidel-Morgenstern, 2005)

Reaction	Separation Separation*	Hybrid Separation	Reactive Separation	Integrated Reaction/
	Distillation	Distillation/ Membrane T.	Reactive Distillation	Reaction Column
Non-Catalytic Catalytic	Absorption	?	Reactive Absorption	?
Homogeneous Heterogeneous Bio-Catalytic	Extraction	Extraction/Reactive Crystallization	Reactive Extraction	Reaction Column
Cell-Lines	Adsorption	?	---	?
Gas Liquid	Chromatography	Membrane Chromatography	---	Reaction Chromatography
Solid Multi-Phase	Membrane Tech.	Membrane Chromatography	---	Membrane Reactor
	Crystallization	Extraction/ Crystallization	---	---

* All combination basically feasible

Figure 106: Reactive and Hybrid Separations

processes.[20] Integration requires more than a sequentially linked and optimized structure. Process integration needs to lead to improved efficiency and reduced capital investment due to simultaneously performed unit operations in one or only a few devices.

Generally, an integration of all reaction types with most separation operations is feasible. This is also valid for several separation or formulation steps combined in one device. Figure 106 illustrates the various integration options. The first column summarizes the most popular reaction technologies. The second column includes widespread separation technologies. Possible combinations of these separation technologies are listed in the third column. Some schemes combine separation technologies sequentially, while others like membrane chromatography perform two separation steps in one device on a micro-scale.

The third column actually shows separation technologies that enhance the separation performance by introducing a reaction step. For example, reactive extraction provides a better equilibrium between the phases due to a scavenging reaction in the receiving phase.

The last column lists truly integrated processes that combine reaction and separation on a molecular level offering efficiency enhancing synergies. Reaction columns and membrane reactors represent the most important integrated reaction-separation devices, while membrane chromatography becomes an increasingly popular, integrated separation technology in biotechnological processes. Integration of solid filtration and reaction is a seldom example for process integration.

---

[20] *Of course, this definition of process integration may be questioned, but it helps to structure the design process.*

Classical Combinations

Figure 107 illustrates a popular design linking reaction and distillation. This design is operated continuously and batch as well. This configuration aims at different

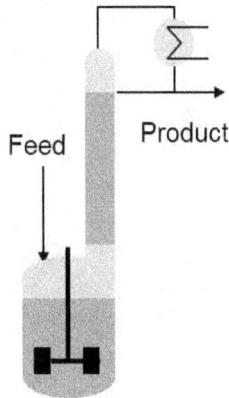

Figure 107: Classical Process "Integration"

goals. The distillation operation may affect the equilibrium in the reactor increasing productivity and product concentration. Removal of a solvent could also initiate crystallization of product or lead to higher product concentration. Different designs are possible. Sometimes vapor from the reactor is not fed to the bottom of the column but as a side feed resulting in a top and bottom product of the column.

Integration is not only possible for reaction and separation unit operations but can

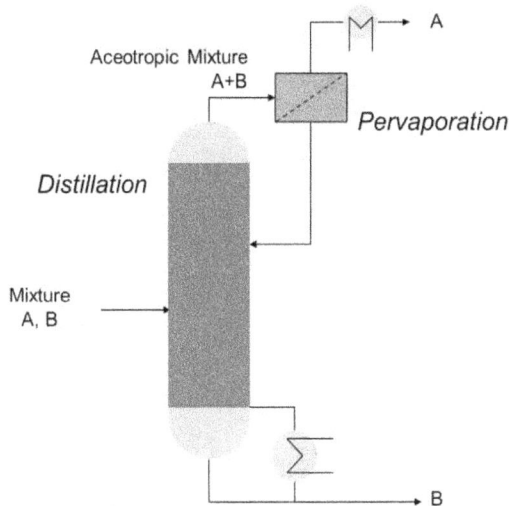

Figure 108: Membrane - Distillation Hybridsystem

also be applied for different separation mechanisms as well. Linking of unit operations has achieved a high level of sophistication during the last years. This type of integration will not be taken into account any longer. The following considerations focus on integration schemes that result in true system synergies.

A combination of distillation and membrane separation can deliver synergetic effects to overcome the challenge caused by azeotropic mixture (for example ethanol/water). Figure 108 describes a combination of distillation and pervaporation where the vapor of the distillation column actually drives a pervaporation unit returning the separated liquid as reflux to the distillation unit.

Reactive separations form an additional integration category. Reactive extraction, for example, uses a reacting agent. This chemical forms a complex with the extracted chemical entity. As a result, the equilibrium is positively shifted increasing driving force and extraction capacity as well. Extractive distillation and absorption function in an analogous way. Reactive separations are widely used, beneficial techniques, but do not really qualify as an integration measure, either.

Dividing Wall Column

Dividing wall columns represent an integrated distillation device combining multiple distillations in one device space resulting in a superior performance.

Normally three component mixtures are separated in pure components using a sequence of distillation columns. Figure 109 schematically describes a possible sequence to separate a mixture containing the components A, B and C. Heuristics once more help to design the distillation sequence. The first column removes component A, while a second distillation column separates the mixture B and C.

A popular alternative uses a side stream to recovery component B reducing the number of columns (Figure 110). The side-stream, however, is always contaminated with A and/or C depending on the location of the side-stream location. The quality requirements for B are decisive, whether this approach is

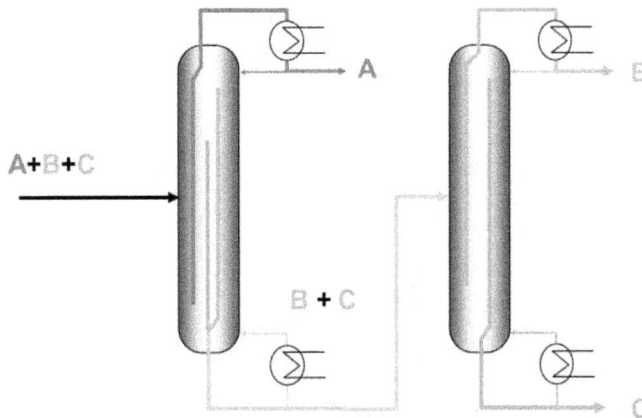

Figure 109: Conventional Distillation Sequence

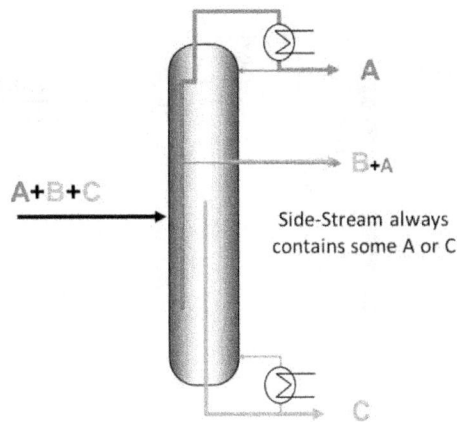

Figure 110: Distillation Column with Side-stream

acceptable. A side-column to purify the side-stream is feasible. This reduces the investment savings, since an evaporator or condenser is not necessary any longer.

The dividing wall column combines the task of two columns into one column by introducing a dividing wall within the distillation column body. Figure 111 gives a schematic description of the main flows of the tree components A, B, and C in a dividing wall column (DWC).

Obviously, components A, B, and C are forced to the top or bottom of the column. The dividing wall blocks the direct path of component B from the column feed to the side removal location of component B. The low boiling component A will move upwards, while C moves downwards within the column.

The middle boiling component B is likely to be found in areas above and below the feeding location of the column. The top section of the column mainly separates component A and B, the lower part separates B from C. This is true before and behind the dividing wall. As a result, the concentration of component B is largest behind the dividing wall. The effect of the dividing wall significantly reduces the likelihood that A contaminates B. Figure 111 qualitatively describes the distribution of the component B within a dividing wall column.

Of course, the quantities of A, B and C have to be appropriate for the fluid dynamics of a dividing wall column. Furthermore boiling temperatures of A, B and C have to be suited as well, since in a dividing column the pressure cannot be chosen separately (compared to a 2-column design).

Dividing wall columns cannot only be applied to multiple distillations, but to extractive distillation as well.

Figure 112 describes the basic set-up of a conventional extractive distillation and a recently proposed design with a dividing wall column. The feed to the dividing wall column contains a mixture of A and B. The component B is moves together with

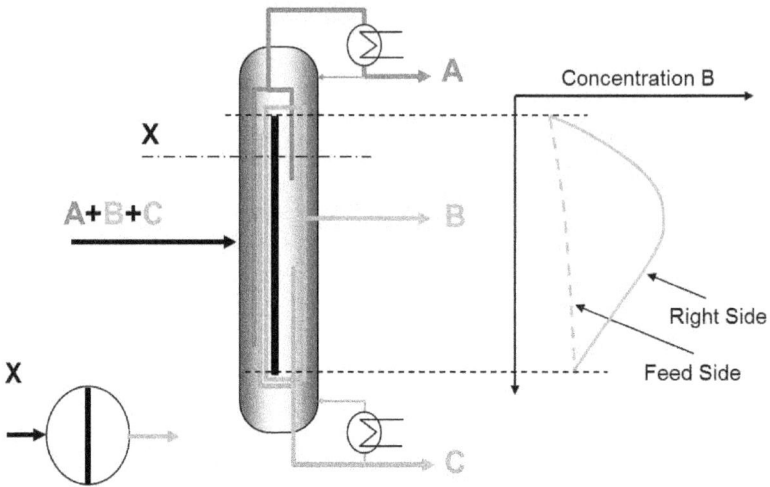

Figure 111: Dividing Wall Column

the entrainer component downwards in the column. Entrainer and component B are separated in the second column.

The dividing wall separates the column areas with component A, B and entrainer from areas with B and entrainer. This configuration recovers component A and B as top products separated by the dividing wall. The entrainer component is removed as bottom product recycled to the top of the column at the feed side of the dividing wall column.

The benefits of a dividing wall column for distillation are

- Significantly lower investment (only 1 condenser, 1 evaporator, 1 column, 1

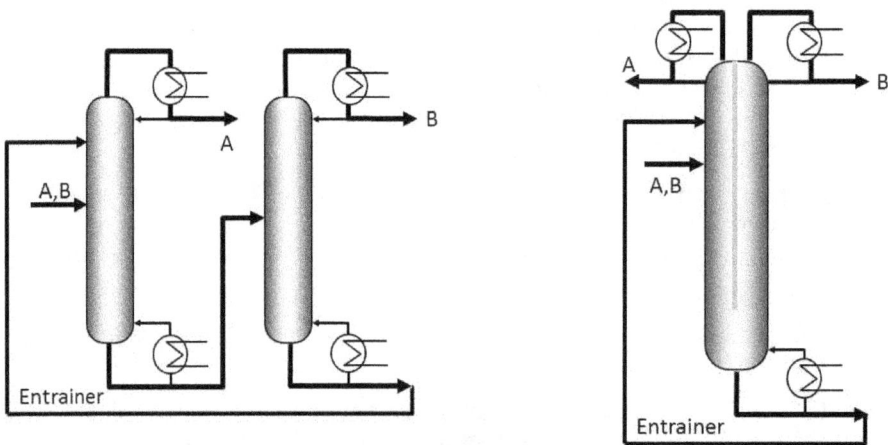

Figure 112: Extractive Distillation with a DWC

column control, less space required)

- Less energy consumption than a 2-Column-sequence (reduced back-mixing, better selectable feed location).
- Less thermal stress resulting by-products

Of course, there are some disadvantages as well:

- Larger size of dividing wall column
- Complex process control
- Larger temperature difference top to bottom (no 2 pressure distillations possible)

There are prerequisites required for dividing wall columns:

- Side stream component >20% of feed
- Separation factors of different fraction in same magnitude preferable

Design Case 11 – Dividing Wall Column for Impurity Removal

The concept of a dividing wall column is demonstrated for the removal of low and high boiling impurities from a final product. In addition, thermal stress generates new impurities in the columns.

Figure 113 shows a 2-column separation sequence. In a one-column approach, the product is removed from the column as a side stream (Figure 114). This product side stream, however, is still contaminated with small quantities of feed and impurities. A side column to purify the product cannot remove low boiling impurities reaching the or newly generated in the side column.

Figure 115 illustrates a dividing column distillation. The product is removed from the reaction column "behind" the dividing wall, while high or low boiling

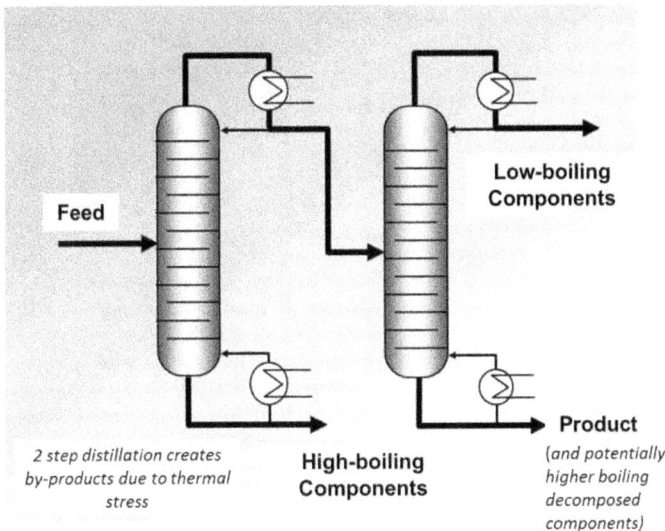

Feed

Low-boiling Components

High-boiling Components

Product
(and potentially higher boiling decomposed components)

2 step distillation creates by-products due to thermal stress

Figure 113: Product Recovery with 2 Column System

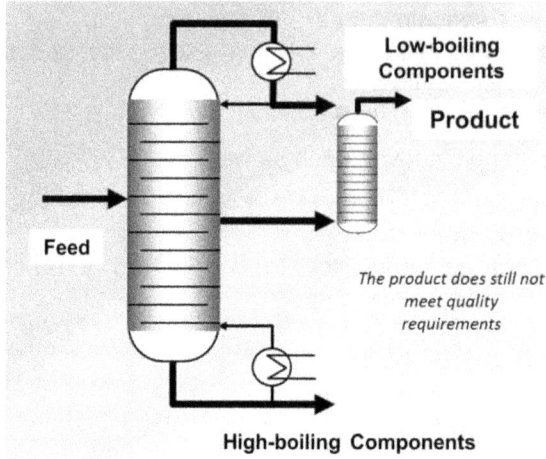

Figure 114: Distillation with Side Column

components are collected at the top and bottom of the distillation column.

The dividing wall enables the removal of a relatively pure product as side stream. There is no back mixing with low-/high- and decomposing by-products due to the flows enforced by the dividing wall.

Thermal stress generating a decomposition by-product prevents this classical 2-column sequence, since the 2-column approach with two evaporators and longer residence time exerts more temperature stress on the heat sensitive product than a dividing wall column.

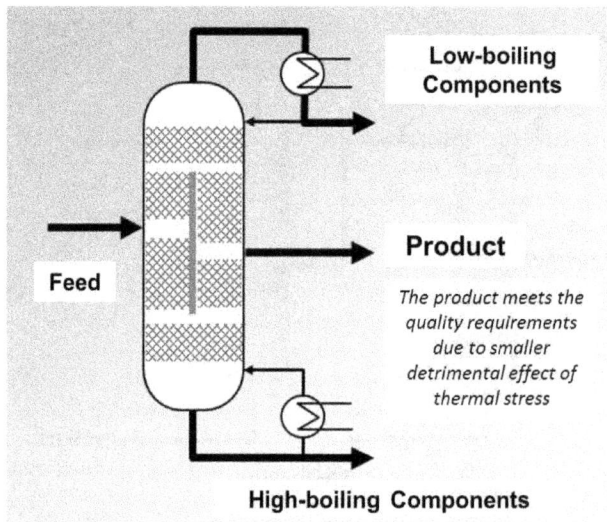

Figure 115: Dividing Wall Column

Reaction Columns

Reaction columns are a concept integrating reaction and separation in one device. The reaction column concept is explained using a catalytic equilibrium reaction converting two reactants A and B into two products C and D. Figure 116 describes a non-integrated, sequential design and a reaction column as well. A fixed bed reactor with a heterogeneous catalyst is used in the standard approach. The reactor effluent consists of a mixture A, B, C, and D that is determined by the equilibrium condition. A sequence of two distillation columns separates the mixture into the products C and D and the raw materials A and B. The configuration of the column sequence depends on the boiling points of A, B, C and. The reactant A and B are recycled to the reactor subsequently converted into the desired products.

Provided the temperature and pressure allow a simultaneous operation of reaction and separation, a reaction column could influence the equilibrium favorably. Figure 116 outlines a reaction column set-up.

The reaction column includes a packing coated with the catalyst. The component A is fed to the reaction column below the catalyst section, while the component B is added above tis section. Component A moves up and component B moves down within the column due to their different boiling points. The catalytic reaction occurs in the catalyst coated packing section of the column. The low boiling reaction product D is stripped out of the reaction section, while the higher boiling product C moves to the bottom of the column. Obviously, the feasibility of a reaction column approach depends on the kinetic and boiling point data of the reaction system.

The final process integration combining the esterification reaction and separation tasks in one device instead of three units is shown in Figure 116.

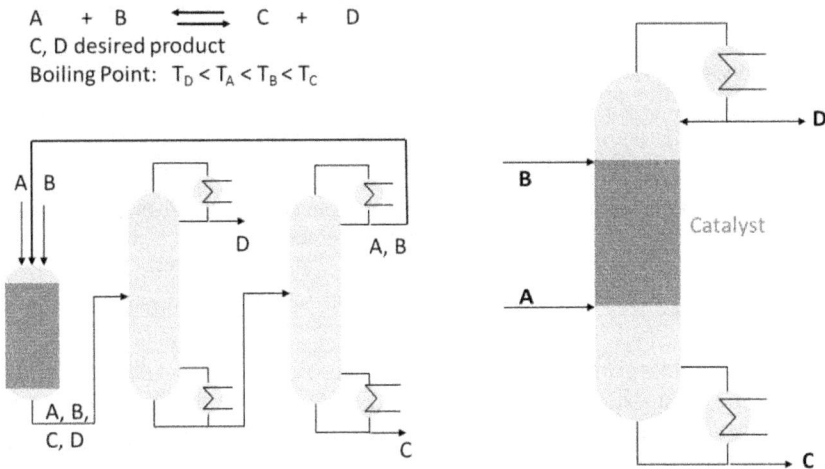

Figure 116: Reaction Column Concept

Reaction Extrusion

Extrusion machines are able to process materials at high viscosities. They are used in many industries. In the plastics industry, they melt granules, add colors or other additives and formulate the intermediate or final product. In the food industry, extrusion technologies provide different consumer products – for example cereals.

Extrusion machine can also perform reactions or separations that occur in highly viscous systems. Extrusion machines frequently accomplish solvent removal from viscous materials.

Twin extrusion machines are the widely applied extrusion concept. These machines are self-cleaning and capable to apply significant forces to materials. The screw elements of an extruder can be modified to seal different extrusion sections from each other. This creates sections in an extruder with tailored-made parameters - to apply more or less stress, to vary temperatures and pressures.

This enables extruders to perform various chemical unit operations such as to melt solids, to mix components, to carry out reactions, to strip solvents and/or formulate products.

Figure 117 describes the basic design of a multi-purpose extrusion unit. In this set-up, classical tasks of a twin-screw extruder is shown for the plastics industry consisting of feeding solids, melting the raw material granules, reacting, dispersing of a filler component, homogenization, solvent removal, and product design.

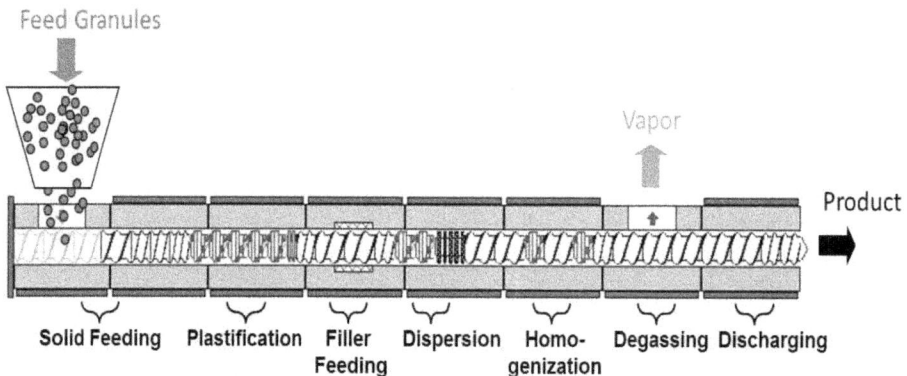

Figure 117: Twin Screw Extrusion for different Unit Operations

Design of an extrusion unit requires testing in laboratory units, although modelling and simulation becomes a valuable tool in recent years (Kohlgrüber, 2008).

In the past, solvent use has often facilitated the handling of chemical systems by reducing the system viscosity. Solvent recovery, however, led to additional energy consumption and capital investment. Extrusion concepts offer process opportunities for innovative, solvent-free, sustainable technology solutions.

Industrial Applications

MTBE (Integration with Reaction Column)

There are various industrial applications of a reaction column. The production of MTBE is an early, well-known example illustrated in Figure 118.

Methanol and Isobutene (in the so-called C4-mixture) is fed to a reaction column. An equilibrium reaction takes place in a catalyst section of the column. The educts are continuously recycled to the reaction section of the column, while the products and the remaining C4-feed/Methanol are steadily been removed from the column. Only two equipment pieces are necessary using a reaction column. Particularly the reaction column enables an operation that favorably influences the equilibrium by continuously removing MTBE from the reaction section.

Figure 118 shows the process with a reaction column on the left reducing the number of equipment pieces. The reaction column requires a sophisticated design highlighted in Figure 118 (on the right).

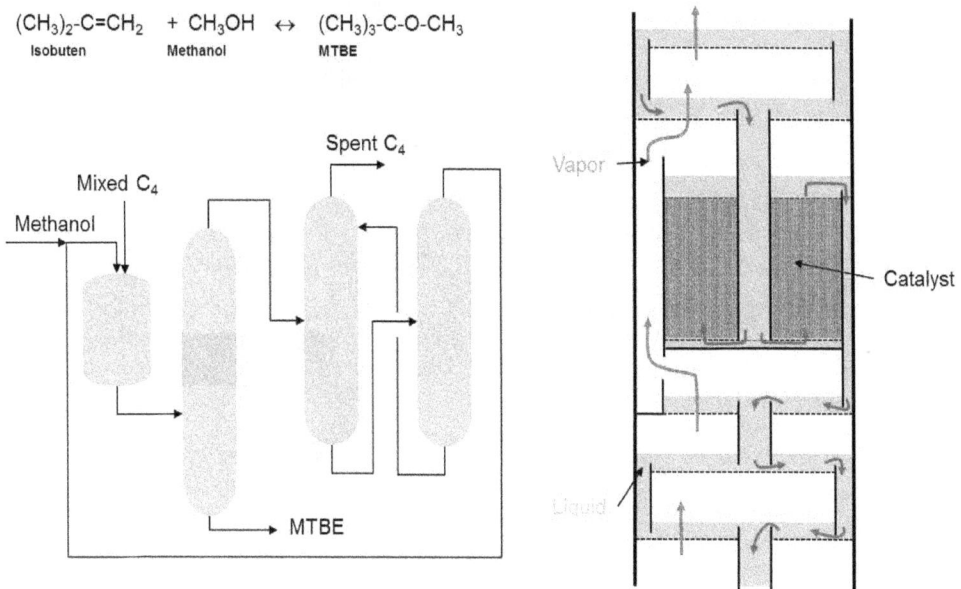

Figure 118: MTBE

Propionic Acid (Integration with Pervaporation)

Gorak gave another example for the production of propyl propionate with reaction column linked to a pervaporation. Chapter 3.2.1 already addressed some features of this approach.

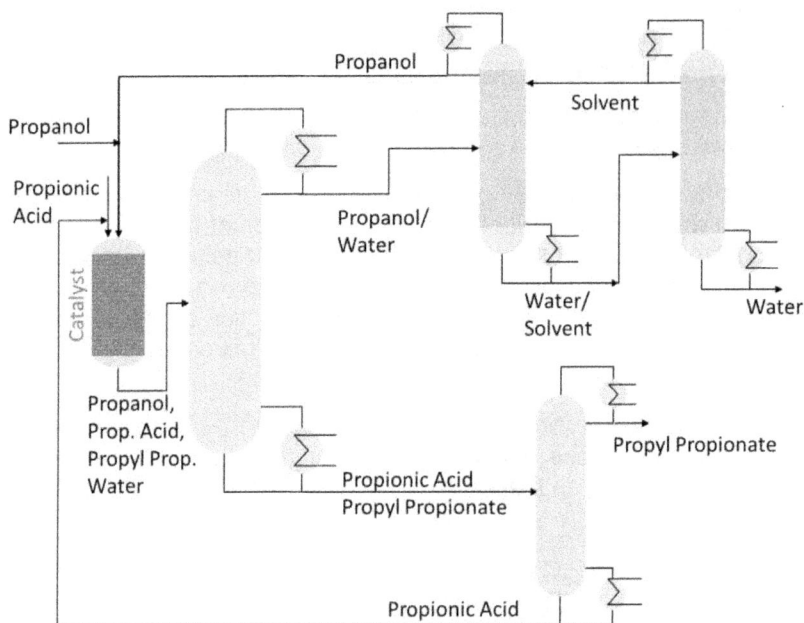

Figure 119: A Standard Design

The standard design would consist of a reactor and a separation sequence. The reaction progresses until the reaction system reaches the equilibrium stage. Then this equilibrium composition is separated in a distillation sequence.

Since propanol/water forms an azeotropic mixture, an appropriate solvent is added to accomplish the final separation. This solvent needs to be recovered in an

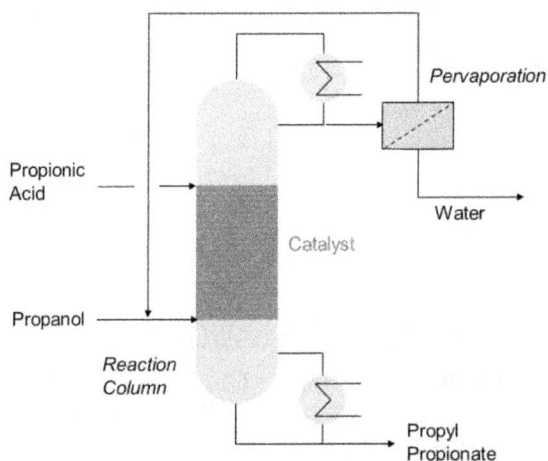

Figure 120: Reaction Column with Membrane Unit

additional distillation column. This design is schematically described in Figure 119.

A reaction column represents an integrated design. Propionic acid and propanol is fed to the reaction column. The reaction takes place in the column section containing the catalyst. An azeotropic mixture of propanol and water is removed as a top product from the column. Propyl propionate can be collected as a bottom product.

The azeotropic mixture of propanol and water represents a special feature of this reaction column. A membrane separation performs the separation of the propanol / water mixture. Water is purged from the process, while propanol is recycled to the reaction column. Chapter 3.2.1 described the process in detail.

Butyl Acetate (Integration with Dividing Wall Column)

The next example describes an application that combines a reaction column with the dividing wall column approach. Here methyl acetate reacts with butanol to generate butyl acetate and methanol. Butyl acetate/butanol and methanol/methyl acetate form an azeotropic mixture.

Figure 121 describes a standard design of a reaction column followed by two columns to recover butyl acetate and methanol, while the two azeotropic mixtures are recycled to the reaction column. This design already is an improvement compared with a reactor separately operated from the distillation column.

Applying the dividing wall column allows a reduction from three to one column. The dividing wall reaction column consists of a catalyst packing on one side of the dividing wall, while the space on the other side of the wall is used to remove

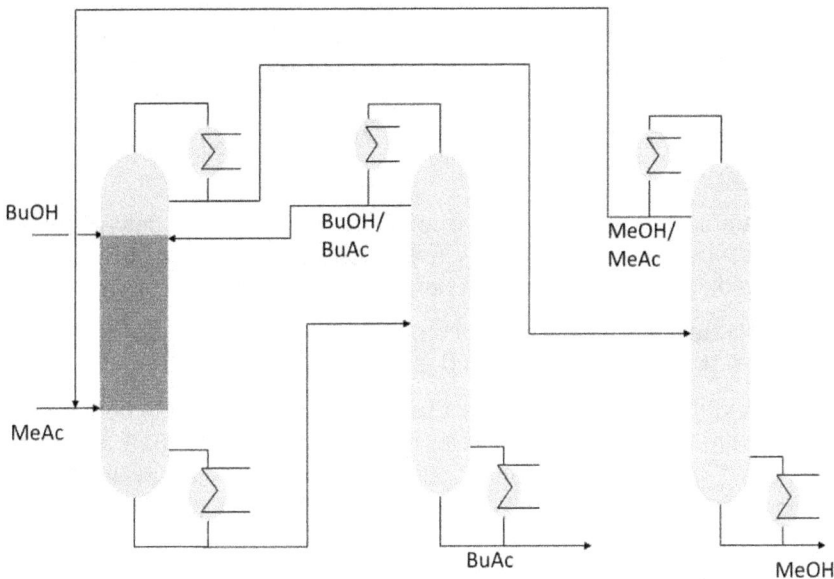

Figure 121: Non-integrated Butyl acetate Process

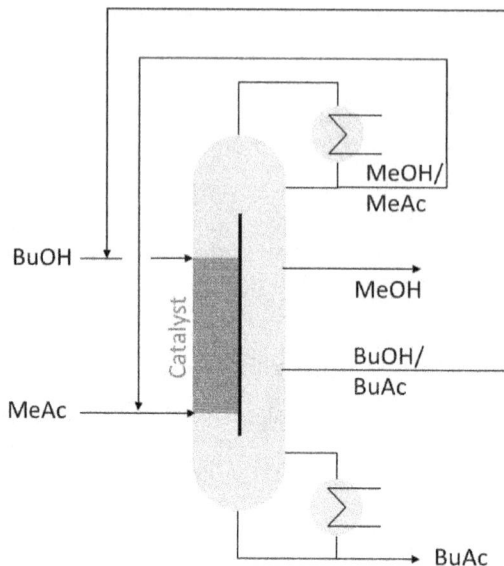

Figure 122: Divided Wall Reaction Column

relatively pure streams of methanol and an azeotropic mixture of BuAc / BuOH. This combination of a reaction column and dividing wall column is illustrated in Figure 122.

The butyl acetate reaction column with a dividing wall represents an excellent example for an integrated process design. Reaction columns combined with a dividing wall are particularly interesting for equilibrium reaction, since capital savings accompany performance improvement as well.

Poly-Silicone

The final example deals with the purification of Si to produce poly-silicon. Figure 123 only describes a section of the purification process. This process links two reactions with different separation and purification operations.

The non-integrated process scheme consists of two reactors and four distillation columns linked by different streams and recycles.

A reaction column combines all unit operations into one equipment piece. The reaction column performs distillations in the upper and lower part of the column, whereas the center of the column includes two reaction sections. Process integration results in significant savings for capital investment and energy consumption as well.

Finally, it is worthwhile to emphasize once more that reaction column and dividing wall column are only feasible, if the kinetic data and material properties for the reaction and distillation operations are suited to apply these integrated concepts.

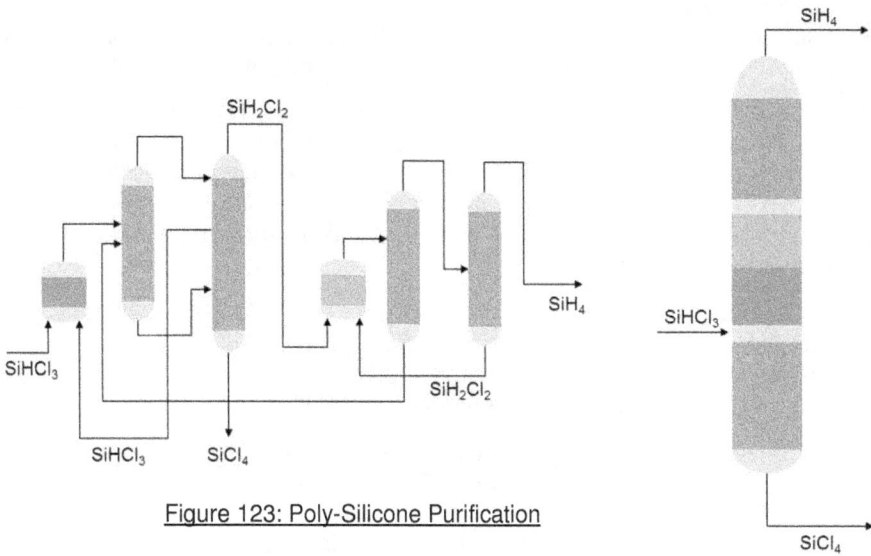

Figure 123: Poly-Silicone Purification

Nevertheless, dividing wall columns have captured a significant share of the distillation column market.

Extrusion Applications

Finally, two examples illustrate the application of extrusion units to process viscous systems into final products.

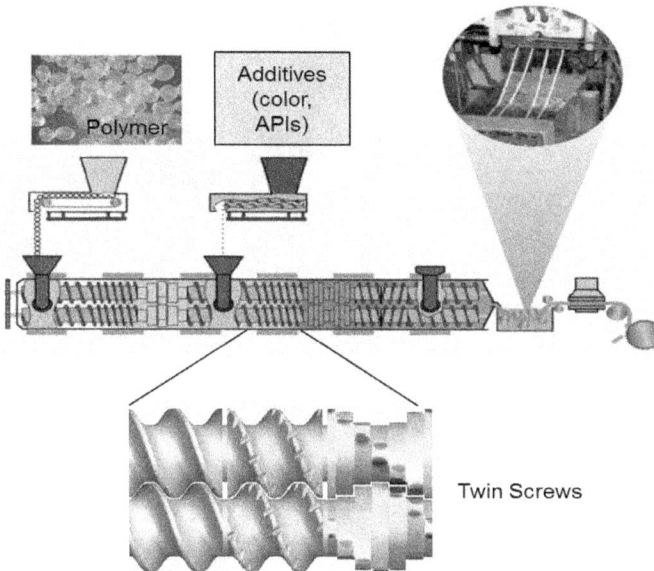

Figure 124: Polymer Filament (Source: Bayer, Leverkusen)

Figure 124 describes an extrusion process to manufacture polymer filament containing active ingredients and/or colors. The extrusion unit consists of sections for polymer melting, color/ingredient addition, homogenization and formulation.

Granular polymer is fed to the melting section of the extrusion unit. Subsequently, an ingredient/color feeder adds powder or liquid components to the melted polymer, mixes, and homogenizes the polymer and additives. The final extrusion section cools down the system and forms fibers using a perforated plate containing appropriate holes. A cooling bath solidifies the fibers subsequently cut and packaged for further processing.

Figure 124 also shows a section of the twin screws. In the past, extrusion design required extensive laboratory testing with expensive extrusion equipment. Today sophisticates simulation software enables computational optimization of screw designs and extrusion operating parameters (Kohlgrüber, 2008).

The production of thermoplastic polyurethanes starting from polyhydric alcohol, butandiol, and MDI is outlined in Figure 124. The extrusion unit performs feeding, mixing, reaction and pelletizing in one device. The extrusion set-up is capable to provide products with a wide range of properties – in particular viscosity.

Reaction-separation extrusion represents a tool applicable to many industries – from food processors and consumer goods to plastics and rubber.

Integrated processes in extrusion machines are generally complex and require special expertise. This expertise is particularly valuable to companies. Many applications are kept as confidential company expertise.

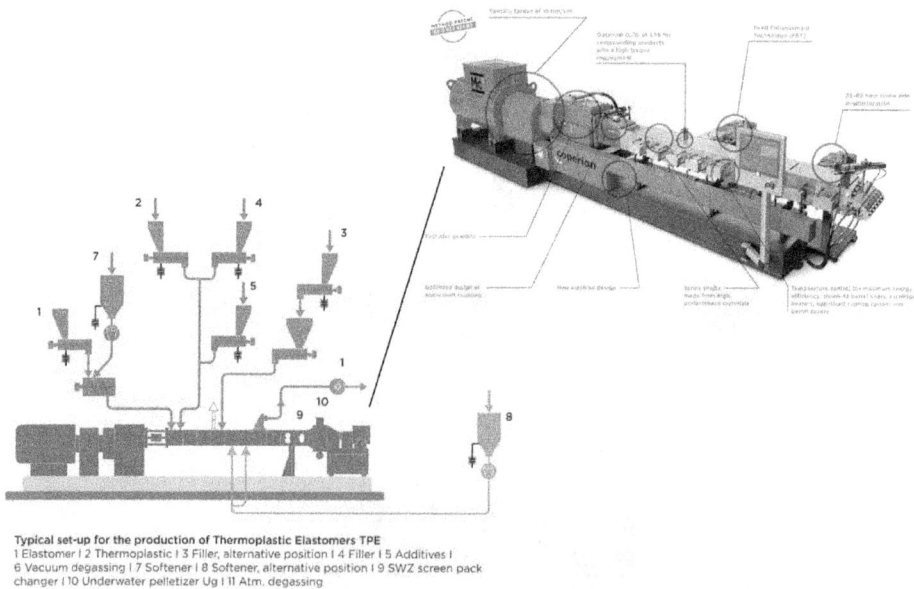

Typical set-up for the production of Thermoplastic Elastomers TPE
1 Elastomer I 2 Thermoplastic I 3 Filler, alternative position I 4 Filler I 5 Additives I
6 Vacuum degassing I 7 Softener I 8 Softener, alternative position I 9 SWZ screen pack
changer I 10 Underwater pelletizer Ug I 11 Atm. degassing

Figure 125: Thermoplastic Elastomers (Source: Coperion GmbH, Stuttgart)

3.4.2 Integration of Heat and Power

The Pinch Point Methodology represents a mature, widely used approach to evaluate and design heat exchanger networks to minimize the energy consumption of chemical plants.

The energy-targeting concept is discussed in detail in chapter 3.2.2. Since there are many literature sources available illustrating the pinch point technology in detail (Smith, 2004) (Klemes, 2011), this chapter focuses on some basic aspects of the design of heat recovery networks. The pinch point methodology provides a comprehensive insight into the energy related features of a chemical process:

- What is the minimal energy consumption?
- Which process areas have a surplus and deficit of energy?
- What is the economically optimal capital investment into heat-recovery equipment?
- How to select the heating and cooling utilities?
- How to integrate reactors, distillation columns, dryer, etc. in the process?
- How to design the best heat exchanger network?

While refineries, power plants or chemical plants in the detailed engineering phase require a complete and systematic application of all elements of the pinch point methodology, the initial design of a heat integrated chemical process has to answer only three questions:

What are the heating/cooling requirements of the basic process? How can the energy relevant unit operations be integrated into the overall process? Which heat exchanger network serves the energy recovery scheme best?

A selection of a pinch temperature ΔT_{Pinch} of 5 to 20°C without any capital targeting is sufficient. A more detailed pinch temperature analysis is only necessary, if the

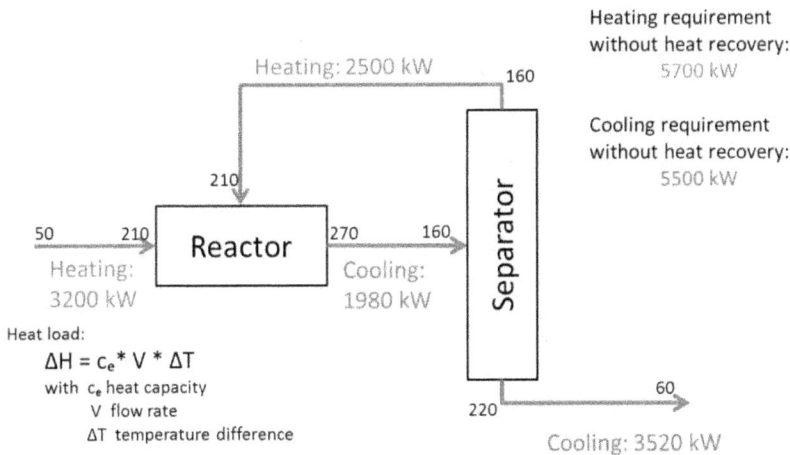

Figure 126: Reactor - Distillation Scheme

"pinch structure" of the process changes within the 5-40°C range.

The generation of a heat exchanger network is demonstrated for the simple reactor-distillation scheme described in Figure 126. The process consists of a reactor and distillation unit linked through a recycle loop. Figure 127 shows the targeting diagram derived earlier.

In this example, the minimum temperature difference for the heat exchange is chosen to 20°C. This leads to a heating requirement of 1000kW and a cooling of 800 kW. The pinch temperature is 170°C. Obviously, a change of the temperature difference to 15°C or 30°C will not affect the characteristic of the pinch diagram. Therefore, it is practical to start with 20°C and evaluate the economic impact of energy savings and capital investment when designing a final heat exchanger network.

Similarly to the calculation of the pinch diagram, there are sophisticated programs available to design a heat exchanger network. A simplified approach, however, is recommended for chemical plants. For this purpose, the hot and cold streams are

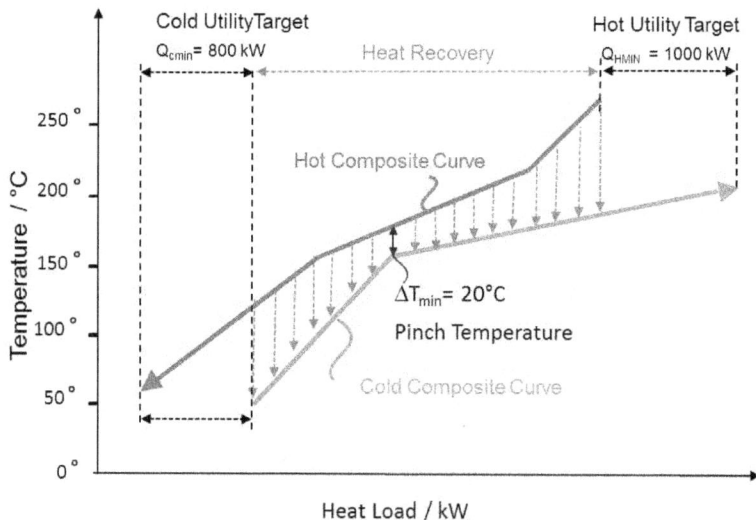

Figure 127: Pinch Point Diagram

lined up in the so-called stream grid (Figure 128). This grid includes information on stream heating/cooling capacities, temperatures and pinch point.

A simplified approach enables the design of an initial design using heuristic rules from chapter 5.

The most important rule separates the design task into two parts – above and below the pinch. An independent heat-exchanger network is generated for each part ensuring that no heat is transferred across the pinch.

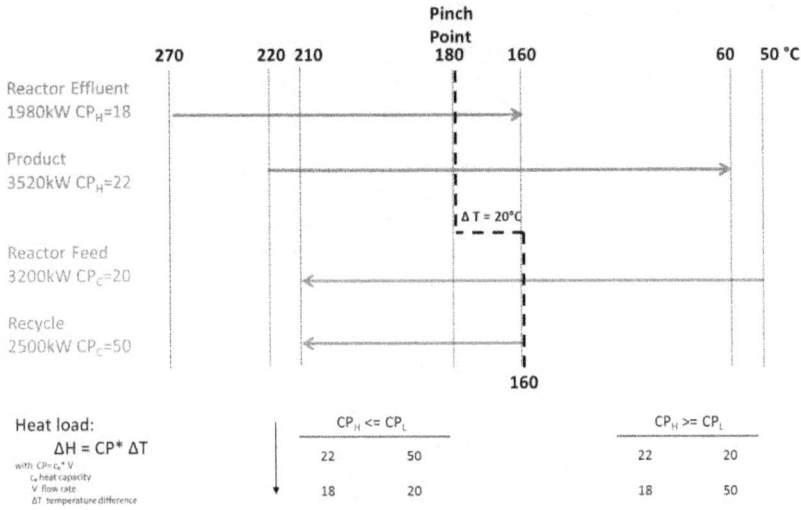

Figure 128: Heat Exchanger Design Grid

The second rule applies the experience that designing the network is most difficult around the pinch area due to the limited driving forces. The CP-rules above and below the pinch directly guarantee that the minimum driving temperature is equal or larger than the selected pinch temperature (here 20°C).

The third rule (sometimes called the tick-off rule) aims at minimizing the heat

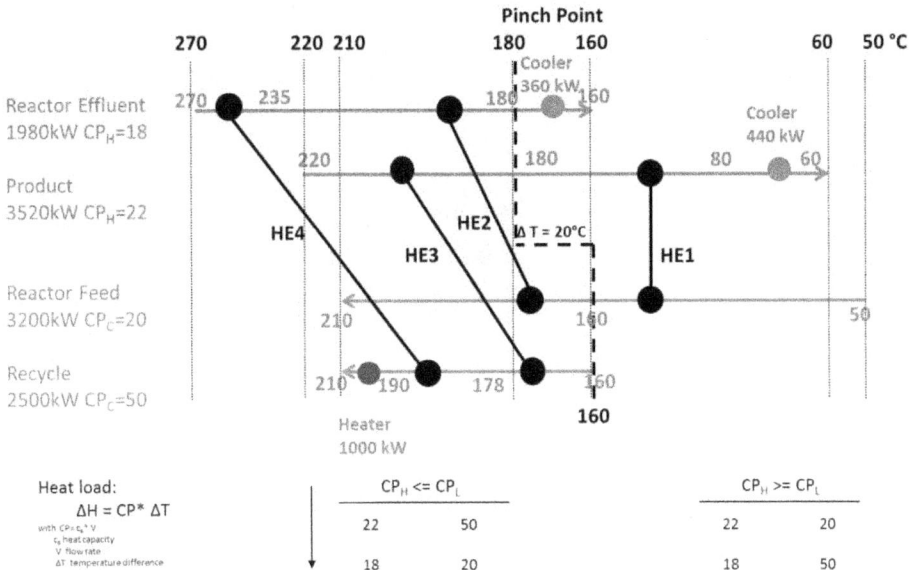

Figure 129: Heat Exchanger Design

exchanger number making them as large as possible.

In Figure 129, the heat exchanger HE 1 is first selected below the pinch. Both coolers follow directly to accomplish the task below the pinch. Applying CP-rule and the tick-off rule above the pinch leads to heat exchanger HE2 followed by exchanger HE3 and HE 4.

In general, several heat exchanger designs could meet the energy target, but differ in capital investment. In chemical processes, operability and safety reasons drive design decision for heat exchanger network (not considerations concerning capital investment). Refineries or power plants, however, require a strict energy and capital targeting and design.

For a simple process, applying these heuristic rules leads straight forward to an acceptable heat exchanger network.

Finally, the overall number of heat exchanger can be estimated using the minimum number of heat exchanger rule. In this case, there are a minimum of seven units assuming there is only one heating and one cooling source. Figure 129 shows the heat exchanger network derived for the reactor-separator process.

Obviously, the selected pinch temperature of 20°C and the selected network is optimal from an energy consumption point of view, but there is no guarantee that the design is economically optimal.

While complex heat exchanger networks such as in refineries can only be evaluated for the best economic solution using design software, chemical plants are simple enough to first design the network with a practical temperature difference (20°C) and then check the individual heat exchanger for economically

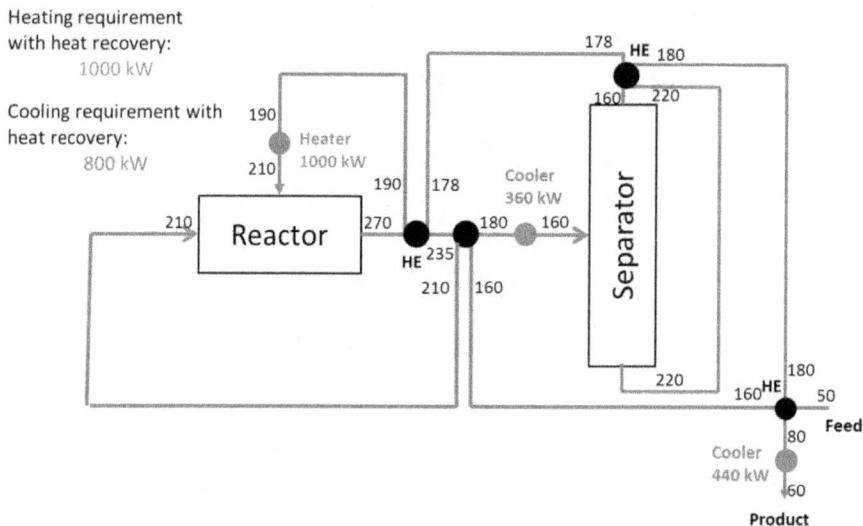

Figure 130: Heat-integrated Process Scheme

reasonable temperature differences.

The heat exchanger network can be translated into a flow sheet describing the unit operations and the respective heat integration. Figure 130 shows this flow sheet. The heat of the separator effluent heats up the feed stream first. Then the hot reactor effluent is used to achieve the desired reactor inlet temperature of 210°. Similarly, the separator effluent is heated up recovering heat from the final product and the reactor effluent. A heater is installed at the separator inlet.

Even for this very simple process, a "random" design of a heat exchanger network is not likely to meet the energy target without knowing the pinch temperature.

Finally, it is also necessary to confirm that this heat exchanger network is operable from a process control viewpoint – during steady-state operation, start-up, and shutdown as well.

Design Case 13: Ethanol Isolation

Finally, the pinch point methodology is applied to a process producing pure ethanol. Since distillation leads to an azeotropic mixture of water and alcohol, a different separation technology has to process this mixture into pure ethanol.

Extractive distillation has been the separation technology in the past. Today membrane technologies or pressure swing adsorption are the technologies of choice.

Figure 131 shows a distillation with three columns linked to a pressure swing adsorption unit. Water removal from a dilute ethanol-water solution consumes a large amount of energy. Ethanol plants, therefore, use a multi-step concentration unit with heat recovery and an adsorption unit requiring less energy to remove the remaining water.

Figure 132 illustrates the pinch point curves of an ethanol purification section. Heat

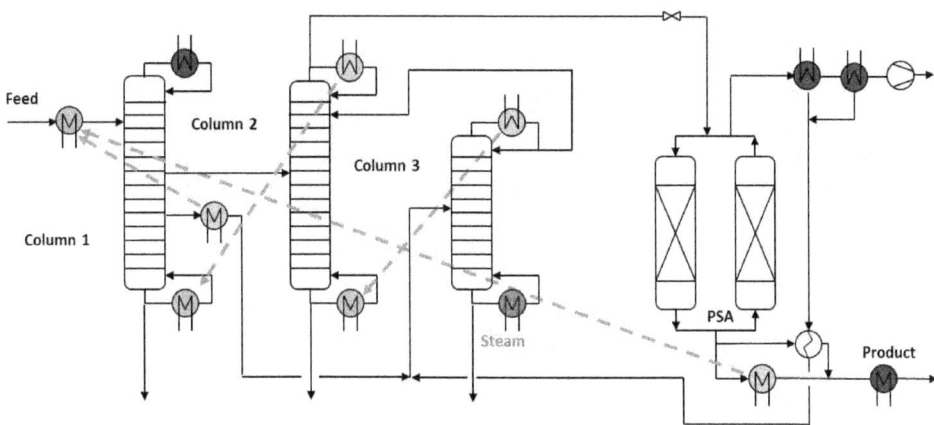

Figure 131: Process Integration for Ethanol Purification

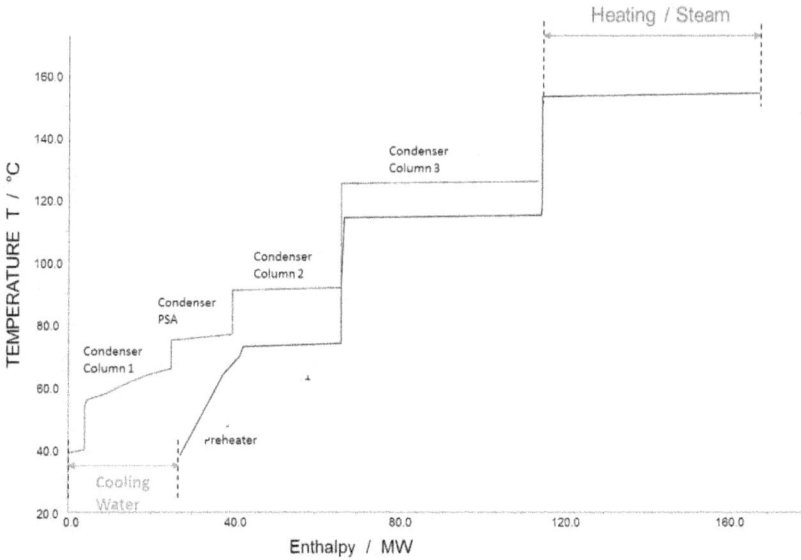

Figure 132: Pinch Point Diagram for Ethanol Purification

"flows" through the column sequence. In this case, the design of the three distillation columns was modified during the pinch analysis to match the column loads optimally.

This could include splitting a big column into two columns. In addition, the column pressures are adjusted to raise or lower temperatures to enable column integration.

A broad overview on process integration is given in various textbooks (Sundermann, Kienle, & Seidel-Morgenstern, 2005), (Klemes, 2011).

There generally occur three shortfalls, when a pinch point study is pursued to reduce the energy cost in a retrofit project:

- Particularly engineers – very familiar with the process to be optimized – carry many convictions, which energy measures need to be pursued. Too much experience can prevent experts from an unbiased look at the process. Engineers new to process are likely to come up with novel, but sometimes, inoperable solutions.
- The framework of a pinch analysis selected only focuses on a process or even process section. It is recommendable to include as many elements as possible initially – start with a look at the total site, before focusing on a single process. Integration requires a wide scope including the power station.
- Economic evaluations consider only current energy and heat exchanger prices. Process designers, however, need to analyze future price development best using simple scenario techniques.

3.5 Process Flexibility – It's about modular Solutions

A technically optimal process design remains the core of an efficient and sustainable chemical production in the future. Process synthesis, analysis, intensification, and integration are important tools to safeguard the quality of a process design.

When the forecast of future product demand becomes more and more difficult due to many product varieties and global competition, capacities of future production processes must be flexibly expandable or reducible.

This constraint adds a new criterion to process design. Flexibility becomes an important feature of an optimal process design. Modular process designs adaptable to fluctuating product demand are enablers of flexible future production systems. The smallest, fully operable process unit defines a basic process module.

Today it is very often more challenging to decide on the future framework the new process has to perform within than to select the best technical design. In other words, in a changing world the capability of a process design to respond to a changing environment becomes an additional criterion, when designing this process.

The Logic of a demand-driven Modularization

As discussed earlier, technology and consumer trends affect the design of processes and plants in various ways:

- Globalization aggravates the forecast of product demand and resource availability
- Individualization requires significantly more flexible processes and plants due to more, smaller, tailor-made products
- Sustainability adds another criterion to process design.

This creates a major problem for manufacturer using large-scale facilities designed for a fixed peak demand. Deviations in market growth or final peak demand result in underutilization or supply problems.

A demand-driven modularization of processes and plants allows an adaption of supply capacities to temporal and local product demand during a product lifecycle.

The design dream for new manufacturing processes of course is a plant that follows these demand profiles as smooth as possible. The process capacity should be easily adjusted to the demand and robust against changes in demand. The individual product demand profiles depend on the industries differing for bulk chemicals, drugs, crop protection, or consumer products.

Classical process designs require an upfront investment in a process based on expected peak demand, while process modules are added in steps following the product demand.

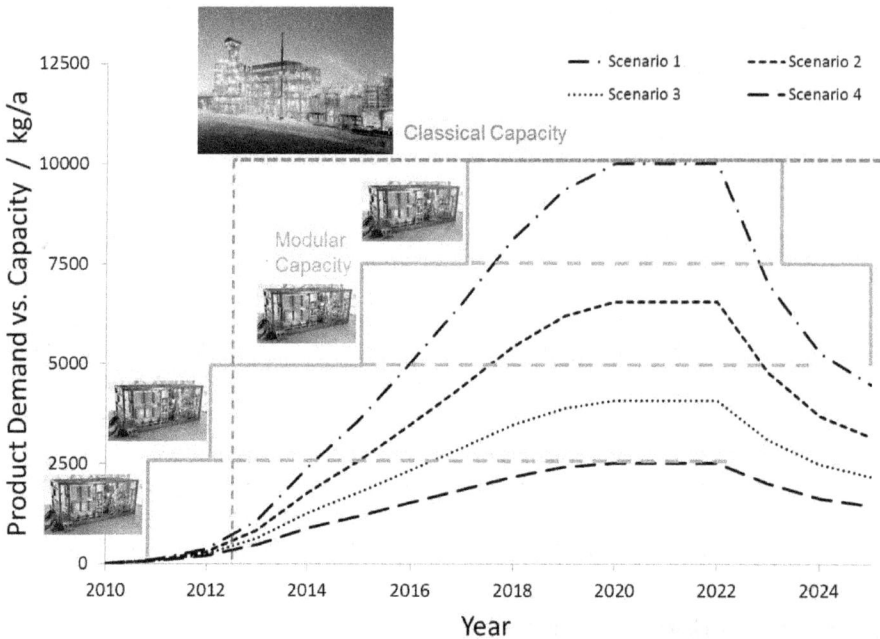

Figure 133: The Demand-Supply Relation (Peak Deviation)

Which process design risks could unexpected product demand create? The demand initially growths slower and/or does not reach the peak product demand predicted in the design scenario.

Figure 133 includes four different demand scenarios to explain a modular expansion strategy. A first module will initially be built with a modular design concept. No additional module is added to the plant, if in contrast to forecast product demand remains low (scenario 4). More modules are added in case of increasing product demand (scenarios 2, 3, 4). An "unplanned" module is implemented, if product demand surprisingly exceeds the peak demand expected initially.

Additionally, modules offer the opportunity for a distributed production. Furthermore, modular plants could be removed at the end of the product cycle.

A large-scale investment cannot be adjusted to the demand curve, if the expected growth does not take place as expected. A classical process already designed for peak demand will never operate fully utilized as long as peak demand does not occur. Manufacturing cost stay above plan endangering profitability and often requiring special depreciations. If product demand exceeds design peak, capacity adjustment for a classically designed process requires a significant effort. A large-scale plant will only meet the design expectations when the design scenario becomes true. In case of a delayed market development, underutilization of a large-scale process occurs during unexpectedly slow demand growth, even if peak sales could finally be reached.

In reality, companies delay an investment in a large-scale process or plant to avoid any risk until the market clearly confirms product demand. Fast acting competitors have often covered the market by then.

A modular production system is particularly beneficial, when

- product demand growths slowly,
- peak sales are difficult to predict,
- local production is desirable.

How does a successful modularization strategy look like? How many modules are optimal? When starts and ends modularization?

Numbering-up and Modularization

A modular process has to supply products during the product life cycle – product development, launch, supply, and withdrawal.

An expansion strategy to overcome the disadvantages of a fixed capacity design has to meet two main technical constraints:

- An expansion of production capacities has to provide product from small laboratory to launch and large production quantities.
- This adaption of process capacities to market demand must not require additional development, qualification, or validation of the capacity to safeguard the product quality.

A capacity expansion strategy preferably covers quantities from a few grams to several thousands of tons.

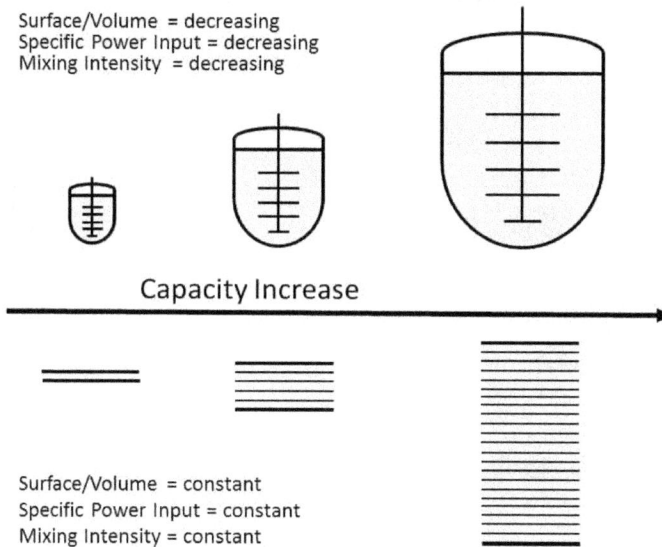

Surface/Volume = decreasing
Specific Power Input = decreasing
Mixing Intensity = decreasing

Capacity Increase

Surface/Volume = constant
Specific Power Input = constant
Mixing Intensity = constant

Figure 134: Numbering-up versus Scale-up

Numbering-up and modularization represent the two fundamental concepts of a flexible capacity expansion strategy. While numbering-up is the expansion approach on an equipment level, modularization is the tool on a process level.

Figure 134 explains the numbering-up concept of a multi-channel reactor in comparison with a classical scale-up approach for stirred tank reactors. Scale-up occurs by numbering-up of identical processing spaces on an equipment level. Capacity expansion takes place without risk, since physical-chemical processes are performed in identical processing spaces.

In contrast, capacity expansion by increasing the stirred tank volume is impossible with identical mixing intensities due to changed dimension of the processing space resulting in additional, expensive process validation (for example new clinical tests for drugs).

Numbering-up offers a major advantage in this example. While fluid dynamics and reaction kinetics are unchanged in the multiplied channels, fluid dynamics in larger stirred tanks cannot match the condition in small lab reactors (for example specific energy input). This can affect selectivity and yield of chemical reactions. Newly created by-products could be detrimental for active ingredients for drugs resulting in expensive, additional clinical testing.

Figure 135 demonstrates the numbering-up approach for a micro-structured reactor. Laboratory tests are pursued in a single channel reactor. Pilot studies take place in a unit with a small number of channels, if piloting is required at all. A large-scale unit consists of several thousands of channels identical to the single channel of the laboratory unit.[21] Numbering-up takes the risk out of capacity expansion, since the reaction mechanisms, flow condition, etc. are always the same in a channels – provided the distribution of material is uniform.[22]

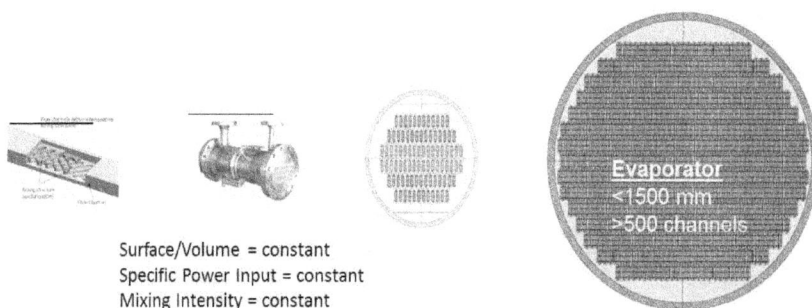

Surface/Volume = constant
Specific Power Input = constant
Mixing Intensity = constant

Figure 135: From Lab to Production by Numbering-up

---

[21] *Kockmann. (2012). CIT, pp. 646-659.*

[22] *Heck. (2012). Verfahrenstechnik, pp. 1-2.*

Figure 137: A basic Process Module (Source: Invite Research Center)

A basic module mainly consists of processing equipment (reactor, separators etc.), piping, and pumps. Sensors and actors to control the process are part of a process

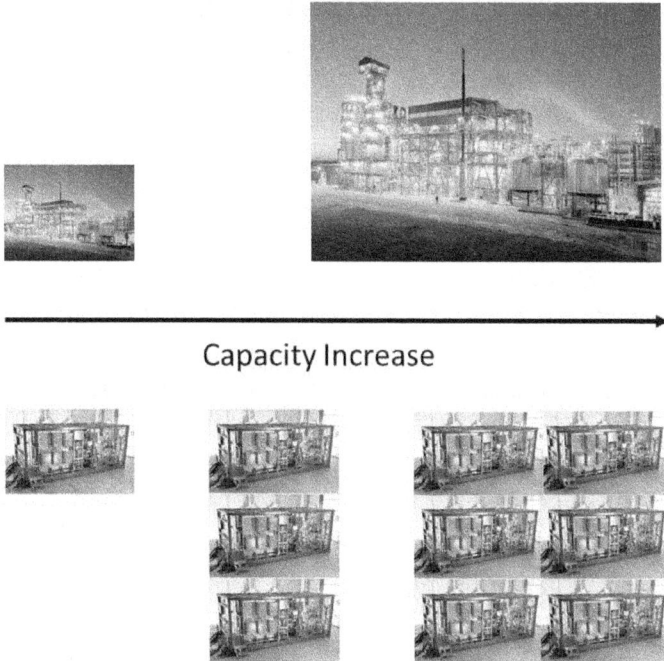

Capacity Increase

Figure 136: Modularization versus Scale-up (Source: Bayer, Leverkusen)

module as well. Utilities or at least connectors to a utility infrastructure complete the set-up of a basic process module. An example of a basic module for a 2-step synthesis including product purification is shown in Figure 136.

Figure 137 compares a modular capacity expansion with a classical large-scale process design. The modular design consists of six modules for peak sales compared to a fixed large-scale plant.

For example, a reaction process consisting of a reactor, heat exchangers, pumps and piping forms a process module. Equipment design of reactor and heat exchangers is based on numbering-up of tubes and channels. The basic module is designed with these equipment pieces linked by piping and pumps. The modularization approach then adds identical modules multiplying the capacity in discrete steps. Since each module is identical to the basic one, this capacity expansion does not create any scale-up issue with respect to product quality. In conventional scale-up, the expansion is accomplished by larger reactors etc. causing the scale-up risk discussed earlier.

The modular approach allows a flexible response to unexpected product demand growth outlined in Figure 133.

Numbering-up and modularization represent a promising approach to address the flexibility design dilemma in times of uncertainty and complexity.

The Flexibility-Cost-Dilemma of modular Designs

Unfortunately, flexibility of modular processes is linked to higher capital investment, since "economies of scale"-effects of classical, large-scale plants exceed the "economies of modularity"-effects of modular plants by far. "Economies of scale"- and "economies of modularity"-paradigms read as follows:

Investment $Cost_{classical} \sim Capacity^m$ Scale-Effect

Investment $Cost_{modular} \sim Module\ Number^m$ Modularity-Effect

"Economies of scale"-exponent m is between 0.5 and 0.9 for classical, chemical-pharmaceutical plants. The "economies of modularity"-exponent is rather between 0.9 and 1 up to now.

What is the impact of these different economies on capital investment and manufacturing cost for classical and modular process designs?

Figure 138 describes the relation between capital investment and plant capacity depending on either scale factor or module number. Both designs rely on the same technologies resulting in equal cost for the identical small-scale plant or basic module.

A capacity increase by a factor of 3 requires 3 modules for a modular plant or larger equipment in a scaled-up plant. Investment cost of a 3 module plant increases by a factor of 3 ($m_{modular}=1$) compared to a factor of 2 for a scaled-up, classical plant ($m_{classical}=0.6$) due to different "economies of scale or modularity".

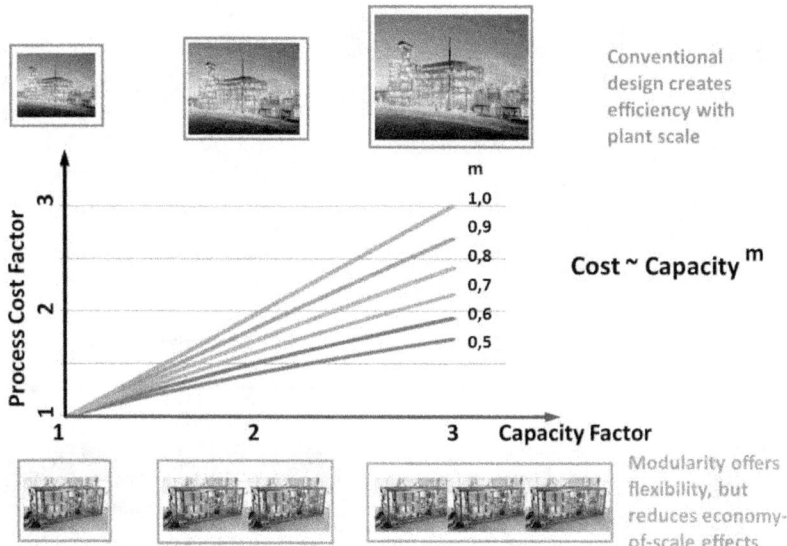

Conventional design creates efficiency with plant scale

Cost ~ Capacity ^{m}

Modularity offers flexibility, but reduces economy-of-scale effects

Figure 138: Capital Investment for Numbering-up vs. Scale-up

Smart modularization and strict standardization can reduce the exponent to 0.7 to 0.9 reducing the capital cost of multi-modular designs. This improved "economies of modularity"-effect results from reduced engineering expenses for the second, third, and following modules. Proper purchase agreements will reduce the cost of equipment units as well provided identical equipment is ordered for the later modules. Nevertheless, an investment cost disadvantage for modular designs remains in comparison to scaled-up designs.

Capital investment affects manufacturing cost primarily through depreciation. Figure 139 describes the impact of higher depreciation on the manufacturing cost. The classically designed large-scale facility reflects the decrease of manufacturing cost at larger scales due to a diluted fixed cost. Modular systems possess almost unaffected manufacturing cost, since depreciation increases with additional modules linearly.

At smaller scales, a modular approach shows advantages. In many cases, a modular approach allows the introduction of novel technologies earlier and with less risk due to the smaller scale than large-scale facilities. Modular concepts create benefits, if a local, decentralized production is desirable. These features result in more flexibility, the disadvantage of higher cost, however, remains in many applications.

Modular production systems offer true "no-regret"-solutions balancing profit maximization with risk minimization.

Obviously, a decision on modular or classical scale-up designs depends on the business area:

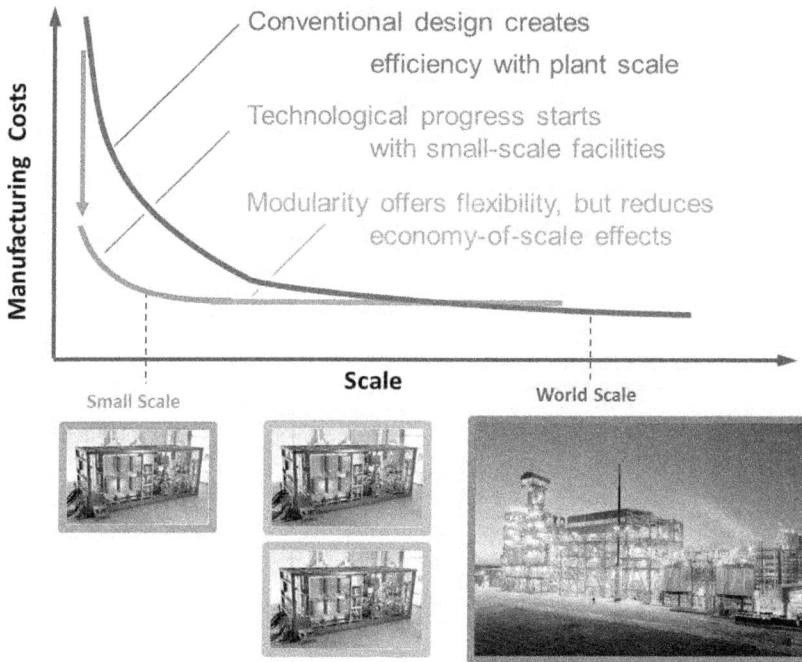

Figure 139: Impact of Modularization on Manufacturing Cost

Large-volume, low-price commodities require minimal capital investment favoring world-scale facilities.

Specialty manufactures with tailor-made, multiple products may be better served with flexible systems.

High-risk producers like drug companies may also prefer risk minimization of modular designs.

Mothes gives a detailed description on the application and economies of modular designs (Mothes, 2015).

3.6 Design Evaluators – It`s about Profitability

Technology and innovation are exciting, but finally profitability decides on the success of a new process design.

Capital Investment totals up the investment necessary to construct a chemical process. *Cost of Goods (CoGs)* characterizes the <u>economic</u> performance of a manufacturing process design.

Annual *Cash Flows* describe the <u>financial</u> performance of a process design. The cash flow summarized for the lifetime of the process and discounted for a present value defines the *Net Present Value* of the process design.

A comparison of net present value and capital investment for the process provides the *Cash Flow Return of Investment (CFRoI)*, a measure for the project profitability.

The economic and financial parameters describe the profitability of process designs (Figure 140).

Since the main purpose of economic evaluations for process design is to select the better solution, the economic models have to focus on parameters necessary to distinguish different designs. Simple estimation methods are often sufficient to compare different alternatives.

A sound investment decision based on project profitability, however, is only possible if input data are complete, quantitative, and accurate.

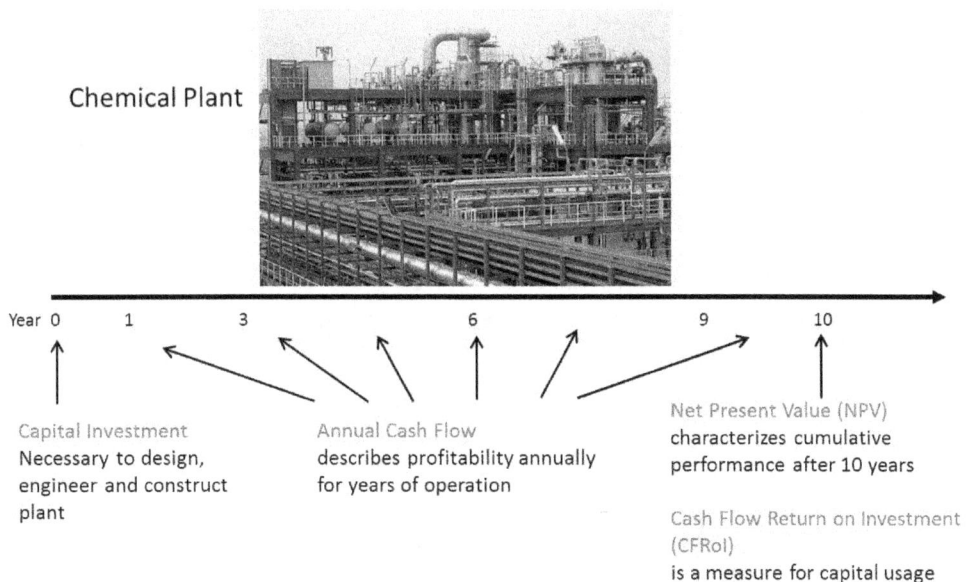

Figure 140: Economic Parameters of a chemical Plant (Source: Bayer, Leverkusen)

Capital Investment

Capital investment generally consists of

- Upfront investment to design, purchase, construct and start-up the equipment and facilities to produce goods
- Working capital (e.g. raw materials, operating materials, inventory)

This capital investment occurs as an upfront investment even before any product is sold.

A detailed description of capital cost estimation is given in the literature (Peters, 2004), (Smith, 2004). Cost Estimation packages are generally part of integrated process simulation software.

In early project stages, the estimates are mainly used to compare process alternatives and select the best design. Absolute accuracy of estimates is less important, the relative accuracy for different designs is important.

Based on the cost of standard equipment, a factor approach is used to estimate the cost of process equipment, process control, piping, building, utilities, and infrastructure for the actual process design. Peters describes this approach in detail (Peters, 2004).

The equipment cost is generally selected for a standard size from a cost database. The actual equipment cost is then adjusted for capacity, pressure, temperature, and material using respective factors M, f_M, f_P, und f_M. Summing up the individual equipment pieces leads to the overall process equipment cost.

Base Equipment: $Cost_{Base}$

Real Equipment: $Cost_{effective} = Cost_{Base} (V / V_{Base})^M f_M f_P f_M$

Process Cost: $Cost_{Process} = \sum f_{installed,i} Cost_{effective,i}$ *for i = 1....n*

$Cost_{Process} = f_{installed} \sum Cost_{effective,i}$ *when* $f_{installed,i} = f_{installed}$

Additional cost for piping, instrumentation & controls, construction, utilities, and engineering is estimated with factors using equipment cost as basis.

$$Cost_{Capital} = + (f_{installed} + f_{erec} + f_{pip} + f_{instr} + f_{Bldg} + f_{util} + f_{log} + f_{eng} + f_{cont}) \sum Cost_{effective,i}$$

Working capital to start-up the plant constructed is linked by a factor approach to the capital investment for the facility. This factor f_{WC} is in the range of 0.1-0.2.

$$Cost_{working capital} = f_{WC} Cost_{capital}$$

Finally, the price inflation is managed by an index I

$$Cost_{capital} = I_t / I_{t=0} (Cost_{capital} + Cost_{working capital})$$

The factors depend on the type of process and plant. The factor approach cannot reach accuracies lower than +/-30%. With design progress, design estimate accuracy continuously is reduced from +/-30% to less than +/-10%.

Exemplary, an overview is given differentiating between a plant that mainly processes fluids and a plant that handles solids to a large extend:

Factors	Fluid Processing f_i	Solid Processing f_i	Range
Equipment delivered	1.	1.0	1.0
Equipment installed	0.4	0.5	0.2 - 0.7
Piping	0.7	0.2	0.2 – 0.8
Instr. & Contr. & Electrical	0.3	0.2	0.2 – 0.4
Building.	0.3	0.3	0.1 – 0.5
Utilities	0.5	0.2	0.2 – 0.6
Logistics	0.2	0.2	0.1 – 0.3
Engineering & Construction	1.0	0.8	0.5 – 1.2
Contingency	0.4	0.3	0.2 – 1.0
Σ	4.8	3.8	4.0

A simple example illustrates the basics of a factor approach for capital cost.

Design Case 14 – Capital Cost Estimation

Figure 141 describes a process with a reactor cascade, distillation column, recycle, and three pumps.

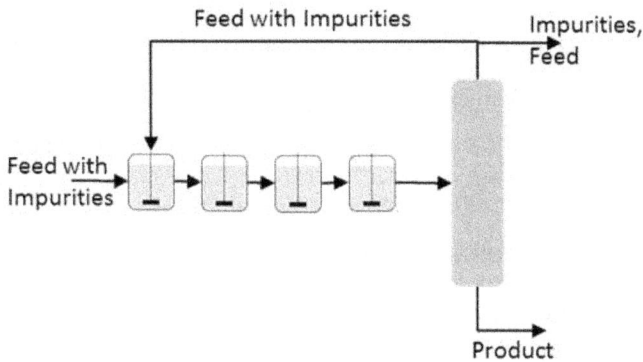

Figure 141: Process Example for Factor Approach

The Equipment list includes:

- 1 Reactor cascade with 4 stirred vessels of 6 m³ (standard material)
- 1 Distillation Column with dia. 0.5 m (standard material)
- Pumps for 100l/min (standard material)

The equipment database provides the following cost data:

Base Reactor Cost: 25000 € V_{base} = 1 m³ (Standard material in 2000)

Scale-factor: M=0.45

Base Distillation Column Cost: 40000 € for dia. 0.1 m (Standard material in 2000)

Scale-factor: M=0.8

Base Pump Cost: 5000 € for 50l/min (Standard material in 2000)

Scale-factor: M=0.6

Inflation Index: $I_{today}/I_{t=2000}$ =2.2   Standard Material: $f_M, f_p, f_T = 1$

Applying an factor approach results in equipment cost for the actual capacities of

$Cost_{reac,effective}$ = $25000*(6/1)^{0.45}*2.2$ = 500 000 €

$Cost_{dist,effective}$ = $40000*(0.5/0.1)^{0.8}*2.2$ = 320 000 €

$Cost_{pumps,effective}$ = $3*5000*(100/50)^{0.6}*2.2$= 50 000 €

$Cost_{process}$ = $1*$ ($Cost_{reac,effective}$ + $Cost_{dist,effective}$ + $3*Cost_{pumps,effective}$) = 2 370 000 €

Assuming a factor of 4 for the completion of the process gives a total investment of

$Cost_{Fixed\ assets}$ = 2 370 000 * 4 = 9 480 000 €

The working capital necessary to start the plant operation adds

$Cost_{working\ capital}$ = $0.10 * Cost_{fixed\ assets}$ with f_{CW} = 0.10

The capital investment totals up to

$Cost_{capital}$ = 9 480 000 € (1 + 0.10) = 10 428 000 €

The accuracy is likely to be less than +-30% for this preliminary cost estimate.[23]

Sales, Operating Result and Cash Flows

The operational performance of the process constructed is the result of the overall business performance. Figure 142 shows a schematic summary simplified description, but sufficient for the process design effort.

The starting point of a profitability calculations are the sales generated with the products manufactured. The manufacturing cost includes all expenses necessary to produce the product such as raw materials, energies, labor and many more. These items can be distinguished in expenses occurring even during idle operation – called fix cost – and expenses only necessary during production – called variable cost.

The difference between sales and manufacturing cost called gross margin represents the first parameter characterizing manufacturing performance. Process yield or energy utilization strongly affects this margin.

---

[23] *The capital estimate example is a fictional case to demonstrate the factor methodology.*

The second cost block covers expenses necessary to run a business. These expenses not directly related to production include expenses for research and development, marketing and sales activities and general management / business expenses. Depreciation is a further item included in this section.

Subtracting these expenses from the gross margin leads to the earnings before income tax (EBIT). Sales, CoGs, Gross Margin, and EBIT are widely used to

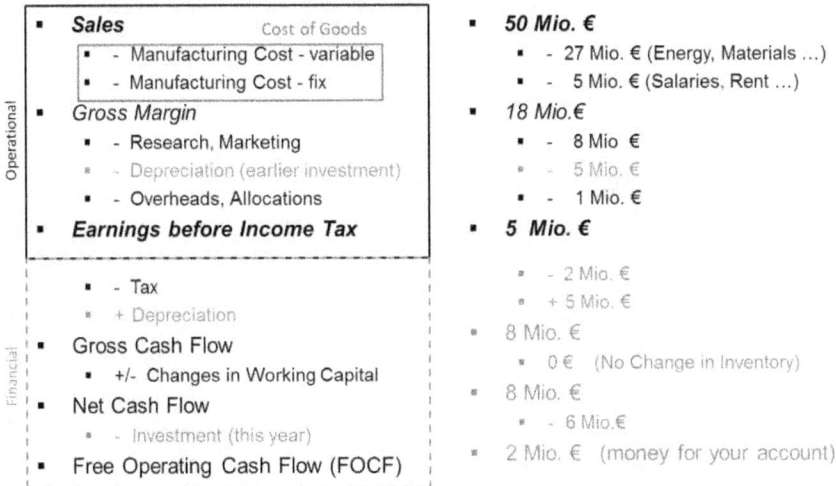

Figure 142: Sales, CoGs, EBIT and Cash Flows

describe the operational performance.

The overall performance, however, requires additional financial figures. The cross cash flow is calculated taken into account taxes and depreciation. Since depreciation is actually an accounting figure (cash was paid already earlier), the real cash flow is determined by adding depreciation back. Some other elements between EBIT and GCF are not considered here, since they are minor and not relevant as long as we focus on process design studies.

Since sales could be accomplished with goods produced earlier (therefore not included in the actual manufacturing expenses) or goods manufactured with cost accounted earlier, but not sold (no sales), inventory changes affect cash flows as well.

Changing inventories due to random fluctuations in sales are not relevant for process design decisions. If two alternate process designs require different inventory levels, the initially necessary inventory investment is already covered by the capital investment.

Cost of Goods (CoGs)

CoGs can also be estimated by a factor approach during early design phases. The process design phase requires a smart approach for estimating cost. Only data should be used relevant to differentiate two optional designs.

In general, raw materials, utilities, operating materials, and logistics are considered variable expenses. Variable manufacturing cost depends on product volumes produced. Depreciation is linked to earlier investment.

During a process design phase, the variable manufacturing expenses are calculated in more detail, while R&D, marketing, sales and administration & overhead are estimated using as % of sales as long as the facility is operated in the range of 80-120% of initial design capacity:

Maintenance: 2-6% of capital investment

Marketing: 1-3 % of sales

Research: 2-10% of sales

Administration: 2% of sales

Depreciation: capital investment / depreciation period

Obviously, process design affects all components of the manufacturing cost. The cost below gross margin is almost not affected by process design. Estimates for these expenses are sufficient.

Figure 142 also includes an example how figures may turn out for a certain accounting period. This arbitrary example is useful to emphasize the relevance of the cash flows. In this example, gross margin, earnings before income tax and gross cash flow are positive.

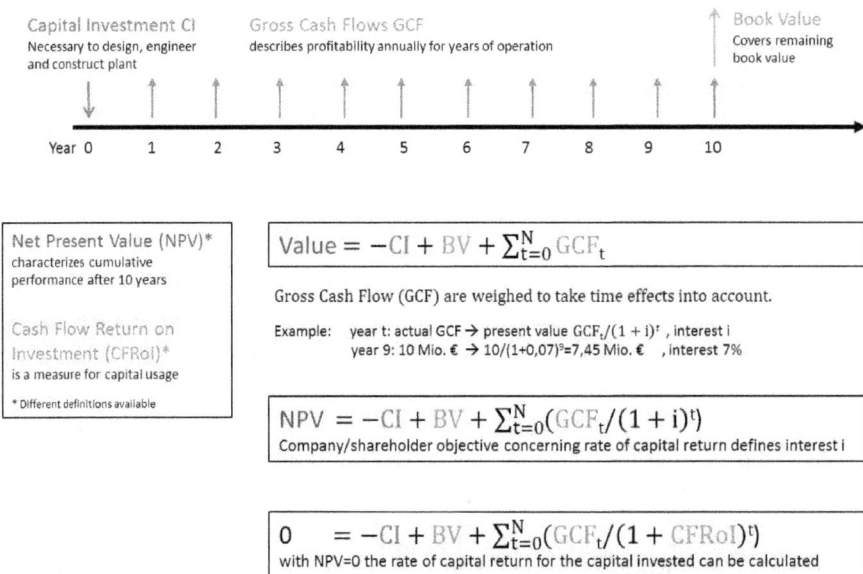

Net Present Value (NPV)*
characterizes cumulative
performance after 10 years

Cash Flow Return on
Investment (CFRoI)*
is a measure for capital usage

* Different definitions available

$$\text{Value} = -CI + BV + \sum_{t=0}^{N} GCF_t$$

Gross Cash Flow (GCF) are weighed to take time effects into account.

Example: year t: actual GCF → present value $GCF_t/(1+i)^t$, interest i
year 9: 10 Mio. € → $10/(1+0,07)^9 = 7,45$ Mio. € , interest 7%

$$NPV = -CI + BV + \sum_{t=0}^{N}(GCF_t/(1+i)^t)$$
Company/shareholder objective concerning rate of capital return defines interest i

$$0 \quad = -CI + BV + \sum_{t=0}^{N}(GCF_t/(1+CFRoI)^t)$$
with NPV=0 the rate of capital return for the capital invested can be calculated

Figure 143: Net Present Value and Cash Flow Return on Investment

Assuming a gross margin of only 12 Mio €/year (due to lower sales prices), the EBIT will become negative (-1 Mio. €/year), the gross cash flow, however, remains positive at 2 Mio. €/year (due to the depreciation effect).

A negative earnings before income tax resulting in a negative cash flow really describes an unsustainable financial performance. The company needs fresh money.

A negative earnings before income tax still resulting in a positive cash flow cannot be accepted for long time, but is often unavoidable in cyclic businesses during downturns.

Net Present Value and Cash Flow Return on Investment

The economic performance of the business activities and the capital invested is characterized by two economic parameters: *Net Present Value* and *Capital Return on Investment*.

Figure 143 illustrates the formulas to calculate NPV and CFRoI for a chemical process plant.

In year 0, the capital investment takes place to construct and start-up the plant operation. For 10 years, the process delivers an annual gross cash flow to be calculated as shown earlier. After 10 years, the process plant possesses a

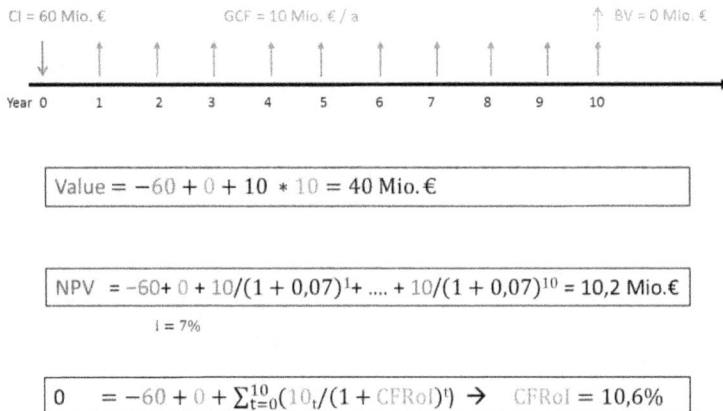

CI = 60 Mio. €   GCF = 10 Mio. €/a   BV = 0 Mio. €

Year 0   1   2   3   4   5   6   7   8   9   10

$$\text{Value} = -60 + 0 + 10 * 10 = 40 \text{ Mio. €}$$

$$\text{NPV} = -60 + 0 + 10/(1 + 0{,}07)^1 + ... + 10/(1 + 0{,}07)^{10} = 10{,}2 \text{ Mio.€}$$
$$i = 7\%$$

$$0 = -60 + 0 + \sum_{t=0}^{10}(10_t/(1 + \text{CFRoI})^t) \rightarrow \quad \text{CFRoI} = 10{,}6\%$$

Figure 144: Example for NPV and CFRoI

remaining book value for the financial balance. Facilities and process plants will be depreciated for a certain period – 10 years for process plants, 25 years for buildings.

The net present value is calculated summing up the initial capital investment (CI), the annual discounted gross cash flows for the accounting period and the remaining book value. Generally the discounted annual gross cash flows are used instead of the nominal cash flows to get today`s value of future earnings. An interest rate is used to calculate the present value from future cash flows.

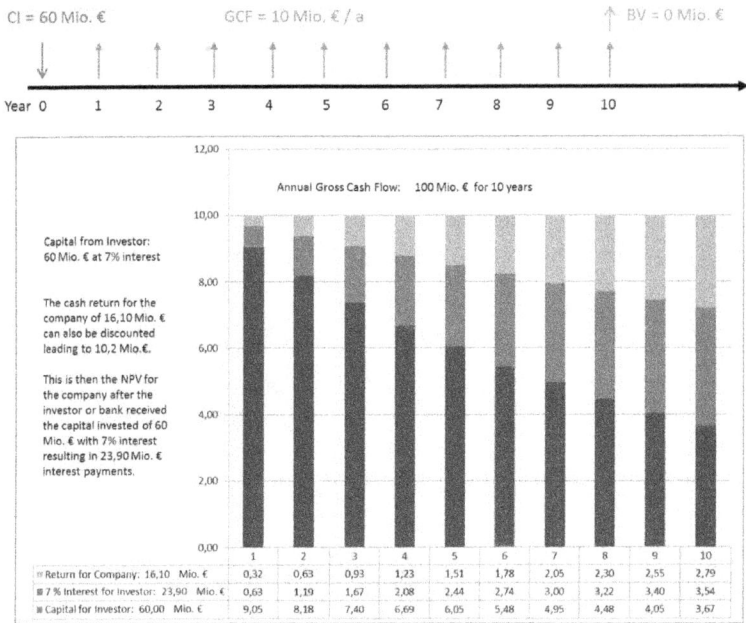

Figure 145: NPV-Curve for Investment Project

Figure 144 gives an example how the mathematics work assuming a CI of 60 Mio €, a BV=0 €, an interest i=0% and 7% and a constant annual gross cash flow of 10 Mio. €/a. The value of the project is 40 Mio € for no interest and 10.2 Mio. € for 7%.

The rationale behind the 7% interest case is that the company/investor would get 7% interest for his investment (the basic hurdle for the company/investor) and an additional profit of 10.2 Mio. €.

The cash flow return on investment CFRoI for the example in Figure 144 amounts to 10.6%. The cash flow return on investment is calculated by search the interest rate for the NPV formula that leads to a NPV=0 at the end of the accounting period (in this example 10 years).

This result could be interpreted an investment comparable to putting 60 Mio. € in a bank account for an interest rate of 10.6%.

Figure 146 gives another interpretation of NPV and CFRoI. An investor provides 60 Mio. € asking for 7% interest. The graph shows capital paybacks. The cumulated paybacks of course sum up to 60 Mio. €. The interest cash amounts to 23.9 Mio. € in 10 years. For example, in year 8 the investor receives 4.95 Mio. € as capital payback. This amount generates an interest of 3 Mio. € for 7 years. In addition, the operation generates a surplus return of 16.2 Mio. € representing a present value of 10.2 Mio. € discounted with 7% for 7 years. Obviously, there are other pay-pack schemes feasible, but this is an illustration of the rationale behind most profitability analysis during process design.

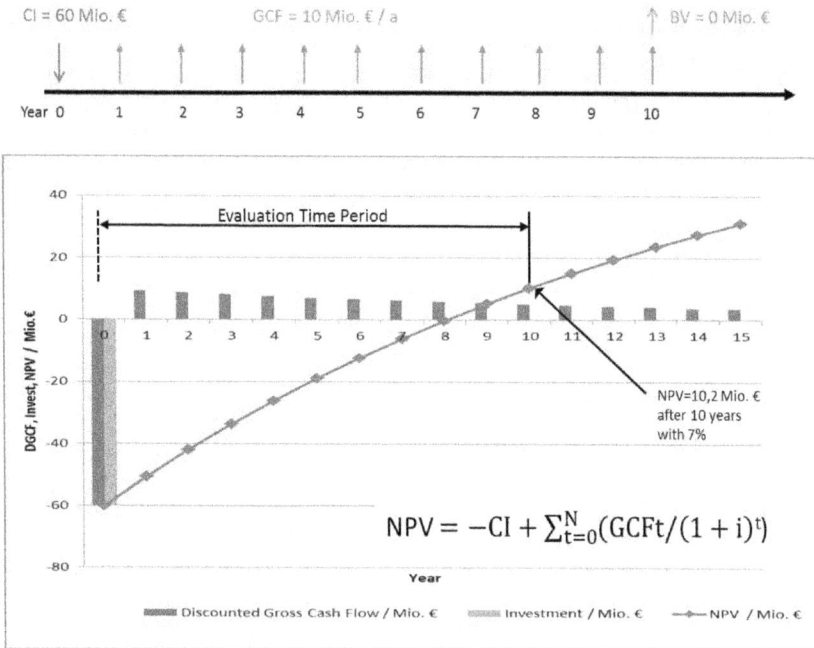

Figure 146: Interpretation of Cash Flows during Investment

The standard description of NPV-curves is shown in Figure 145. The one-time bar of year 0 represents the capital investment. The repeating bars indicate the discounted annual cash flows due to investment and operation. In year 8, the NPV reaches 0. The NPV after year 10 becomes 10.2 Mio. € as the curve (cumulated discounted cash flows from the beginning) indicates.

The desired outcome of a profitability evaluation could be summarized for different design alternatives as follows:

- NPV $_{design\ 1}$ describes the financial revenues of the selected design that should be positive (and higher than competing designs NPV $_{design\ 2,\ 3...}$).
- CFRoI $_{design\ 1}$ of the selected design has to overcome CFRoI $_{hurdle}$ and to be higher than competing designs CFRoI $_{design\ 2,\ 3}$.

4 Design Case Studies

This chapter demonstrates the design methodology. Figure 147 gives a summary of the methodology outlined in chapter 2 and specified in more details in chapter 3. The methodology includes a technical and an economic procedure. Technically heuristic rules provide a guideline to select the best design in a systematic approach. Economic features are evaluated applying economic potential ideas. The technical design heuristics are summarized in the appendix.

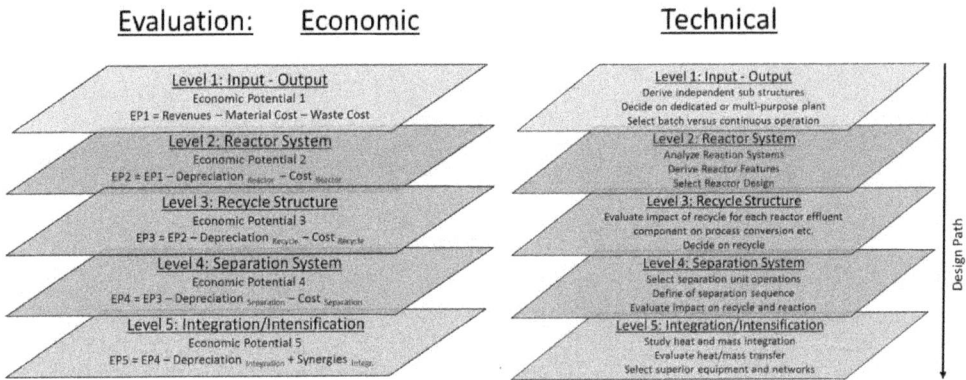

Figure 147: Design Methodology

In reality, a design task does not follow the design methodology "slavishly". Design steps vary for grass root design or retrofitting. Design tasks rely on differing data on kinetics or material properties – sometimes only qualitative, in other cases quantitative.

4.1 Technical Design

The technical design methodology is illustrated for a grass root design of a new integrated chemical site. Only qualitative data are available. Goal of the design task is a preliminary process flow sheet.

Input Data

The design task is to develop a preliminary design for a plant site consisting of a network of processes. The site manufactures five large-volume products P1, P2, P3, P4, and P5. There are 10 educts required and 4 intermediates produced.

Figure 148 summarizes the chemical pathways to synthesize these five products. In addition to the chemical pathways, some relevant qualitative data are listed necessary to make first design decision.

These data include information on the reaction kinetics and material properties. The kinetic data indicate a variety of reaction systems – from multi-phase, competing schemes to high viscous systems. The boiling points are given for components that are liquid at ambient air pressure.

Although these data are only qualitative, heuristic rules can provide initial design decisions. Of course, some decisions on the technical design need to be evaluated by detailed quantitative simulations and experimental laboratory or pilot testing. The outcome of these evaluations may alter the process design.

The products and their chemical synthesis pathways selected allow a demonstration of the key design steps outlined earlier. Nevertheless, this design case mirrors a broad range of chemical processes performed in many chemical companies.

For example, the reaction pathway leading to the first intermediate ZP1 is characteristic for many gas-liquid reactions. Hydrations of basic chemicals belong to this process class. Reactions with chlorine or ammonia gas with benzene or toluene important for many plastic products form a similar challenge. The reaction system resulting in product P1 includes some aspects well known from isocyanate production.

The reaction pathway leading to intermediate ZP4 is representative for transesterification reaction relevant for basic chemicals, biodiesel and many chemical intermediates – particularly important for integrated chemical sites. The reaction system producing intermediate ZP5 frequently occurs in plants creating agro-chemical or pharmaceutical intermediates or final active ingredients.

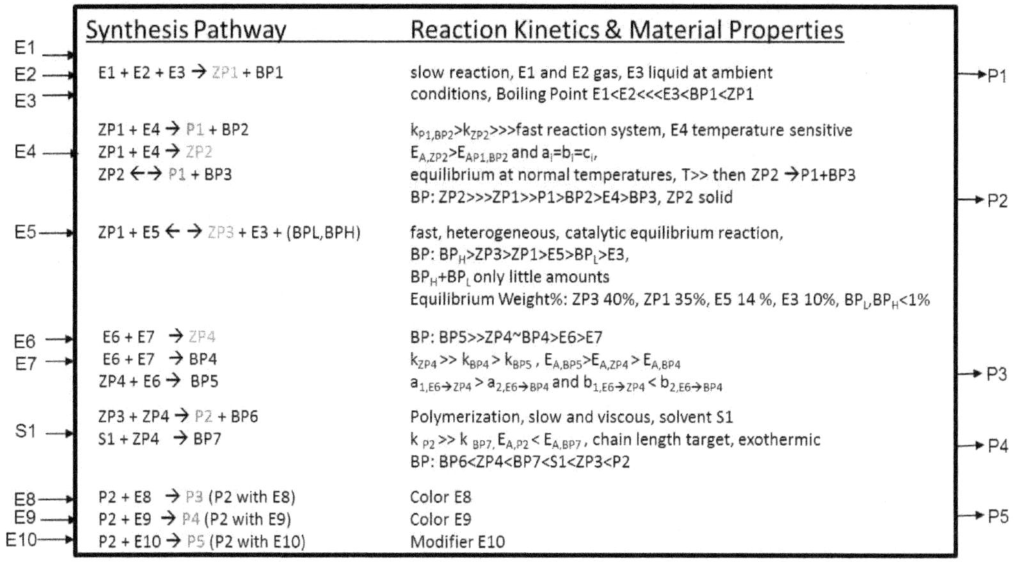

Synthesis Pathway	Reaction Kinetics & Material Properties
E1	
E2 → E1 + E2 + E3 → ZP1 + BP1	slow reaction, E1 and E2 gas, E3 liquid at ambient conditions, Boiling Point E1<E2<<<E3<BP1<ZP1 → P1
E3	
ZP1 + E4 → P1 + BP2	$k_{P1,BP2}>k_{ZP2}>>>$fast reaction system, E4 temperature sensitive
E4 → ZP1 + E4 → ZP2	$E_{A,ZP2}>E_{AP1,BP2}$ and $a_i=b_i=c_i$,
ZP2 ←→ P1 + BP3	equilibrium at normal temperatures, T>> then ZP2 →P1+BP3
	BP: ZP2>>>ZP1>>P1>BP2>E4>BP3, ZP2 solid → P2
E5 → ZP1 + E5 ←→ ZP3 + E3 + (BPL,BPH)	fast, heterogeneous, catalytic equilibrium reaction,
	BP: BP_H>ZP3>ZP1>E5>BP_L>E3,
	BP_H+BP_L only little amounts
	Equilibrium Weight%: ZP3 40%, ZP1 35%, E5 14 %, E3 10%, BP_L,BP_H<1%
E6 → E6 + E7 → ZP4	BP: BP5>>ZP4~BP4>E6>E7
E7 → E6 + E7 → BP4	k_{ZP4} >> k_{BP4} > k_{BP5} , $E_{A,BP5}$>$E_{A,ZP4}$ > $E_{A,BP4}$
ZP4 + E6 → BP5	$a_{1,E6\to ZP4}$ > $a_{2,E6\to BP4}$ and $b_{1,E6\to ZP4}$ < $b_{2,E6\to BP4}$ → P3
ZP3 + ZP4 → P2 + BP6	Polymerization, slow and viscous, solvent S1
S1 → S1 + ZP4 → BP7	k_{P2} >> k_{BP7}, $E_{A,P2}$ < $E_{A,BP7}$, chain length target, exothermic
	BP: BP6<ZP4<BP7<S1<ZP3<P2 → P4
E8 → P2 + E8 → P3 (P2 with E8)	Color E8
E9 → P2 + E9 → P4 (P2 with E9)	Color E9
E10 → P2 + E10 → P5 (P2 with E10)	Modifier E10 → P5

Figure 148: Input on Reaction Pathways and Material Properties

Finally, product P2 represents a viscous polymer such as high performance polymers. Product 2 is modified adding colors or polymer linkers to produce formulated products P3, P4, and P5 representative for polymers and coatings.

Input-Output Structure

The technical design follows a hierarchical approach as outlined in Figure 147.

The first level decision focuses on the input-output structure. Based on the input data of Figure 148, the reaction pathways are used to derive a simple structure of reaction, separation, and formulation steps linking the different reaction systems. Figure 149 shows this flow sheet selected in a straightforward procedure.

A process designer needs to answer two questions at the *Input-Output Level*:

- Do we want to establish dedicated processes or a multi-purpose plant?
- Do we operate a continuous or batch operation?
- Can we create subsystems subsequently treated independently?

Heuristic rules are used to evaluate a dedicated versus multi-product concept (level 1: rule 1-8 in chapter 5).

A first look at the reaction systems indicates that tailor-made reactor structures and parameters optimize selectivity and yield favoring a dedicated approach (rule 2). Products are large volumes favorably produced in dedicated structures (rule 3). Since design has to provide a flow sheet for an integrated site, dedicated processes enable better integration options (rule 4).

Since there is no information available that requires a multi-product design, we design first the optimal dedicated process. After completing the design, it is recommended to compare the process equipment with respect to overlaps enabling a multi-purpose approach.

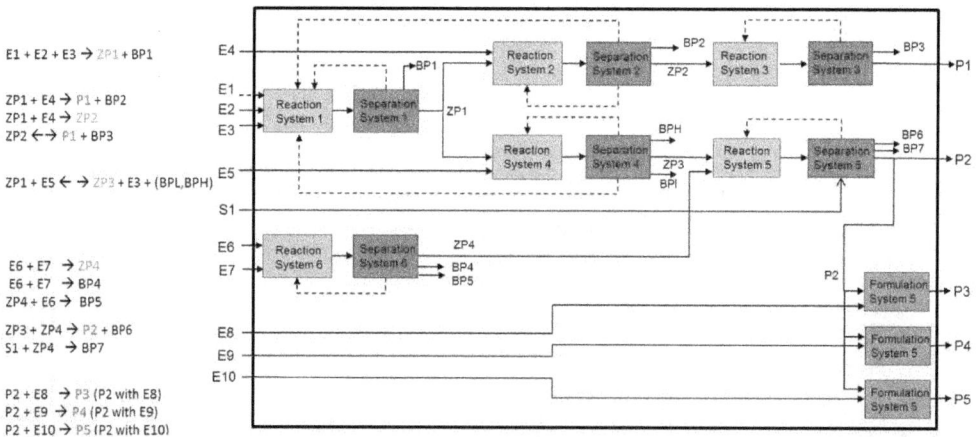

Figure 149: Preliminary Flow-sheet Structure

Heuristics are also used to decide on a batch or continuous operation (level 1: rule 9-20 in chapter 5).

In general, a decision for dedicated processes anticipates a selection of continuous operation (rule 13). Many reasons for a dedicated process are valid for a continuous operation as well.

A continuous operation offers more opportunities for tailor-made optimization. Generally, large volume products are best suited for continuous, dedicated process designs. Generally, it is easier to switch from continuous to batch operation than from batch to continuous operation.

In a next step, the designer analyzes the complete structure in Figure 149 for subsystems with unique features. The reaction pathways in Figure 149 offer a maximum of seven systems consisting of reactors, separators, and/or formulators. If the processes are synthesized for the complete structure simultaneously, a very complex problem needs to be solved.

To divide the complete structure into subsystems facilitates the design task for the smaller subsystems. Of course, these separately designed processes must not lead to a sub-optimal overall plant design.

An approach to select "independent" sub-systems is to study whether a single processing device can perform different process steps.

All boxes in Figure 149 could be separately performed leading to a maximal number of equipment pieces. Alternately, all educts could be added to one reactor, the various reactions take place and the reactor effluent components are then separated into products, intermediates, and educts. Intermediates and educts are recycled. Obviously, this approach minimizes equipment demand, but does not allow for the optimal selectivity, yield, and productivity of each reaction system.

A first screen of Figure 149 recommends the following sub-systems:

- Reaction system 1 is designed as a separate system. Different products use the intermediate ZP1 as a raw material (more flexibility – rule 20). Reaction 1 includes 2 phases (gas-liquid) justifying a separate sub system (rule 16).[24]
- Reaction system 2 and 3 appear to include synergies, if performed in one sub system simultaneously (rule 15).
- Reaction 4 is an equilibrium reaction requiring special designs to maximize selectivity and yield (rule 16).[25] Reaction 5 is a high viscosity system requiring a special reactor. A joint rector for reaction 4 and 5 appears to be unlikely. Furthermore, availability of ZP4 may be more flexibly handled, if systems 4 and 5 are addressed in different subsystems (rule 20).
- Reaction system 6 is a complex system that is best optimized individually (rule 16 and 17).

---

[24] *Actually the "reverse" version of this heuristic is used.*

[25] *Decisions for dedicated processes often indicate a necessity of more sub systems reflected in heuristics as well.*

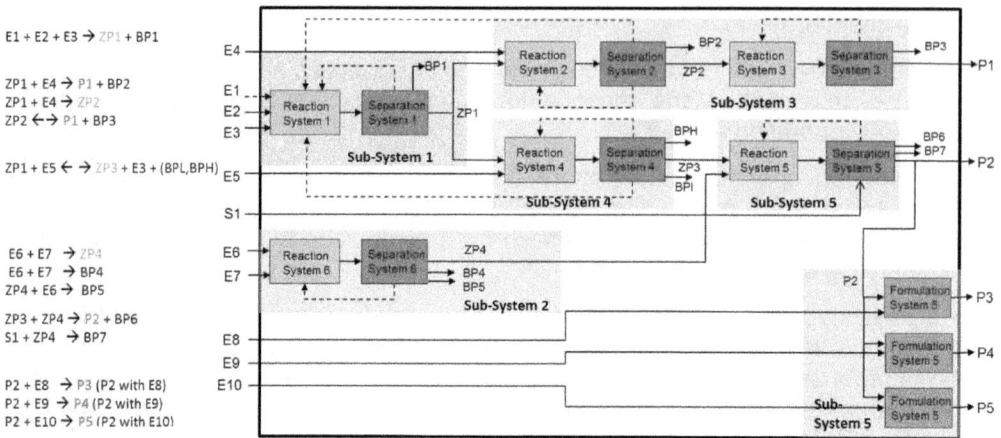

Figure 150: Selection of Sub-Systems (Level 1)

- A joint design for formulation system 7 offers more flexibility (rule 20).

The next issue concerns the necessity for a special separation unit for each reaction sub system. Actually, the earlier decision on sub systems reduces the separation issue to the separation system 2 and 3 for by-products BP2 and BP3, excess educts E4 and ZP1, intermediate ZP2 and product P1. Since synergies cannot be excluded, when reactions and separations 2 and 3 are performed simultaneously, the design of sub-system 3 aims at combining separation 2 and 3 (Rule 15).

Figure 150 shows six subsystems to be synthesized in independent design tasks.

Design of Sub-System 1

The design of sub-system 1 starts outlining a reactor. Figure 151Figure 148 shows the reaction system 1. The reaction is a gas-liquid system. Kinetic data indicate a slow reaction without competing or consecutive reactions. An inexpensive stirred tank reactor is selected to perform this reaction. The reactor effluent is likely to contain all educts and the product – actually an intermediate converted in the next process unit – and a by-product BP1. Figure 151 gives a first scheme of the reactor system.

The next hierarchical level addresses the recycle system. Based on the heuristic rules,   the product ZP1 and the by-product BP1 are separated, while the educts E1, E2, and E3 are candidates for a recycle or purge. Furthermore, one of the raw materials E1, E2, and E3 could also be added in excess (Figure 152).

Since E1 and E2 are gaseous components at ambient pressure, the reactor needs to be operated at higher pressure to support transfer of gases E1 or E2 into the liquid phase E3. This, however, offers an opportunity to recover E1 and E2 in a flash-type separation step easily.

E1 + E2 + E3 → ZP1 + BP1

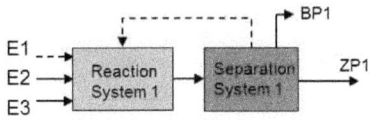

Slow reaction,
E1/E2 gas and E3 liquid at ambient conditions
Boiling Point E1<E1<<<E3<BP1<ZP1

2-phase system:
→ mass transfer E1 and E2 limiting?

Slow Reaction:
→ no issue with selectivity, maximize productivity
→ effluent with E1,E2,E3,BP1,ZP1

Reactor Selection:
→ Inexpensive bubble column or stirred tank

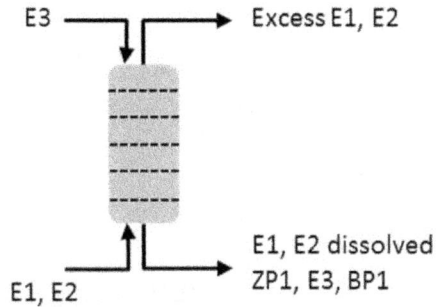

Figure 151: Sub System 1 – Reaction (Level 2)

The reactor could be a bubble column with a continuous liquid phase of E3 at higher pressure or a droplet reactor with a gaseous, continuous phase of E1 and E2 at ambient air. A decision can be postponed to a later design phase when an economic optimization becomes feasible.

Figure 153 describes a separation sequence. In a first step, a flash unit enables a separation of gaseous E1 and E2 from ZP1, BP1, and E3. A distillation column

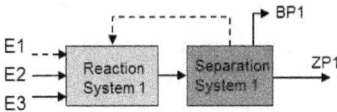

E1 + E2 + E3 → ZP1 + BP1

Slow reaction,
E1/E2 gas and E3 liquid at ambient conditions
Boiling Point E1<E2<<<E3<BP1<ZP1

Slow Reaction:
→ effluent with E1,E2,E3,BP1,ZP1
→ E1, E2, E3   recycle or purge?
→ No recycle of ZP1 and BP1

Recycle Selection:
→ Recycle E1, E2 with gas compressor
→ Recycle liquid E3
→ Separate ZP1, BP1

Any feed component in excess?

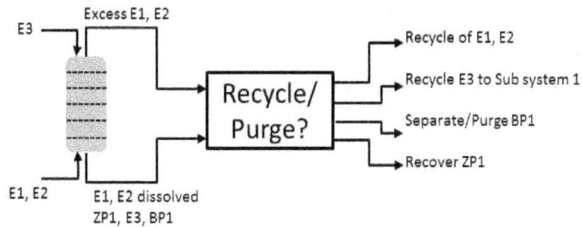

Figure 152: Sub System 1 – Recycle (Level 3)

E1 ---
E2 → | Reaction System 1 | → | Separation System 1 | → ZP1
E3 →

↑----→ BP1

E1 + E2 + E3 → ZP1 + BP1

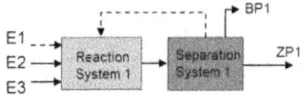

Slow reaction,
E1/E2 gas and E3 liquid at ambient conditions
Boiling Point E1<E2<<<E3<BP1<ZP1

Separation Sequence:
→ Gas-liquid flash separation (E1, E2 from E3, ZP1, BP1)
→ Subsequent recovery of ZP1 from E3 and BP1
→ Finally separation of ZP1 and BP1 (purge?)

Separation technology Selection:
→ Inexpensive flash drum
→ Recycle with gas compressor
→ Distillation columns to separate E3, ZP1, BP1

Any feed component in excess?

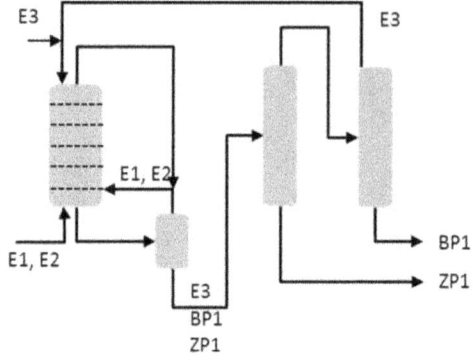

E3 ... E1, E2 ... E1, E2 ... E3 ... E3 BP1 ZP1 ... BP1 ZP1

Figure 153: Sub System 1 – Separation (Level 4)

then separates ZP1 from E3 and BP1. If economic and technical aspects allow, a purge solution could replace the distillation column separating E3 from BP1.

This sub-system 1 represents a candidate for process intensification (level 5). Obviously, the 2-phase reaction system might be mass transfer controlled and, therefore, is always a candidate for process intensification. The logic of smaller bubbles, higher pressures, and more efficient bubble separation in a centrifugal field applies to this processing step (Figure 154).

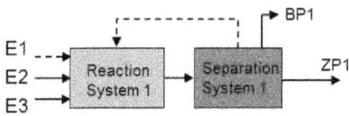

E1 ---
E2 → | Reaction System 1 | → | Separation System 1 | → ZP1
E3 →

↑----→ BP1

E1 + E2 + E3 → ZP1 + BP1

Slow reaction,
E1/E2 gas and E3 liquid at ambient conditions
Boiling Point E1<E2<<<E3<BP1<ZP1

2-phase system:
→ mass transfer E1 and E2 limiting?
→ Increase transfer area and driving force

Intensification:
→ Operate at highest pressure possible
→ Intensify shear stress for smaller bubbles
→ Try internal packing/sieves
→ Evaluate reactor centrifuge

Integration:
→ Evaluate integrated column
→ Test dividing wall column

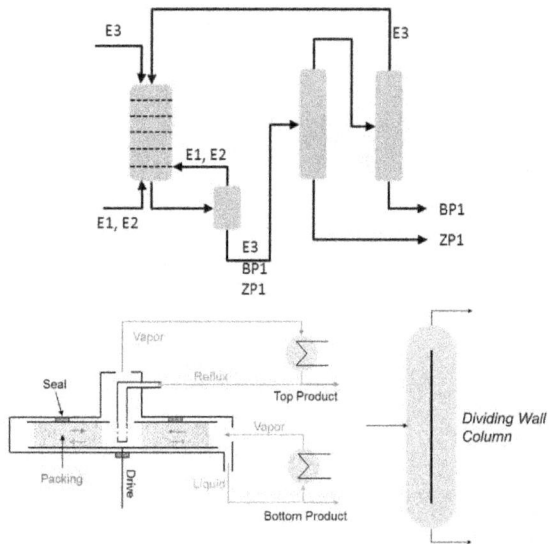

E3 ... E1, E2 ... E1, E2 ... E3 BP1 ZP1 ... BP1 ZP1

Vapor ... Seal ... Reflux ... Top Product ... Vapor ... Dividing Wall Column ... Packing ... Drive ... Liquid ... Bottom Product

Figure 154: Sub System 1 – Intensification/Integration (Level 5)

Design of Sub-System 2

Figure 155 gives a summary on the sub system 2. This sub-system consists of competing and consecutive reactions characterized by a qualitative description of kinetic data (activation energies, concentration orders). A classical analysis of this reaction system directly leads to cascade of stirred tank reactors with an excess of E7 and a gradual addition of E6 to keep the concentration of E6 low.

E6 + E7 → ZP4
E6 + E7 → BP4
ZP4 + E6 → BP5

Competing, consecutive reaction,
BP: BP5>>ZP4~BP4>E6>E7
$k_{ZP4} \gg k_{BP4} > k_{BP5}$, $E_{A,BP5} > E_{A,ZP4} > E_{A,BP4}$
$a_{1,E6 \to ZP4} > a_{2,E6 \to BP4}$ and $b_{1,E6 \to ZP4} < b_{2,E6 \to BP4}$

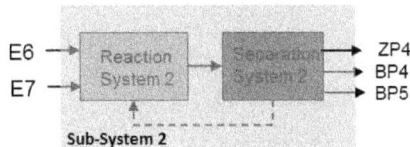

Reactor Operating Features:
→ Low concentration E6
→ High concentration E7
→ Temperature profile

Reactor Design:
→ Reactor cascade
→ Semi-batch Operation
→ Excess E7

Separation:
→ Extractive Distillation for BP8/ZP5

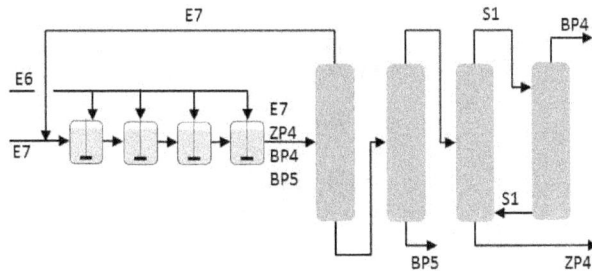

Figure 155: Sub-System 2 – Summary

The temperature level should be higher at the first reactor and reduced from reactor to reactor to maximize the selectivity towards product ZP4 desired. Such a reactor scheme is also recommended by the heuristic rules summarized in the Appendix.

Since the boiling points of ZP4 and BP4 are very similar, distillation is not the separation technique preferable. An extraction solution could overcome this limitation provided a suited solvent is available. A potential separation sequence is derived in Figure 155 using the heuristic approach. There are alternative solutions feasible.

Design of Sub-System 3

Although the sub-system analysis of level 1 suggested a joint evaluation of reaction and separation system as sub system 3 to utilize synergies and minimize the number of equipment pieces, both systems are first treated separately to generate insight in the process later used to optimize overall sub system 3.

Since the rate towards P1 is favored at lower temperatures, an optimal reactor is operated at low temperature for this reaction system. Low temperature leads to less ZP2. Low temperature, however, also shifts the equilibrium towards the solid intermediate ZP2. ZP2 formed once cannot be recovered at low temperatures completely.

A solid removal device could recover ZP2. Subsequent heat treatment favors a conversion of ZP2 towards P1 (and BP3).

A 2-reactor system consisting of a first reactor at low temperature operated with an excess of ZP1 and a subsequent reactor at very high temperature to shift ZP2 to P1 provides a high conversion towards P1.

Since the competing reactions are very fast, reactions could be run to completion within a reactor with reasonable size – particularly if one component is fed to the reactor in excess.

Since E4 is temperature sensitive, an excess of ZP1 could be used to completely consume E4 resulting in an effluent of the first reactor likely to consist of ZP1, BP2, P1 and solid ZP2.

A hydro-cyclone or filter could be used to recover ZP2 that then is treated at high temperature to be converted to P1 and BP3. Neither P1 nor BP3 need to be

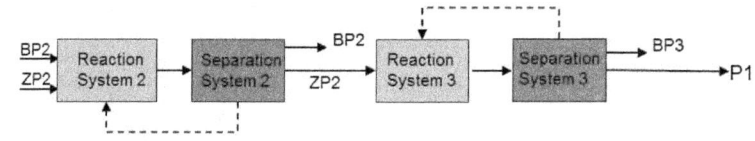

ZP1 + E4 → P1 + BP2
ZP1 + E4 → ZP2
ZP2 ↔ P1 + BP3

$k_{P1,BP2}>k_{ZP2}>>>$fast reaction system, E4 temperature sensitive
$E_{A,ZP2}>E_{AP1,BP2}$ and $a_i=b_i=c_i$,
equilibrium at normal temperatures, T>> then ZP2 →P1+BP3
BP: ZP2>>>ZP1>>P1>BP2>E4>BP3, ZP2 solid

Fast reactions 1 and 2:
→ Mixing essential
→ Excess of ZP1 to consume E4 totally
→ Low temperature favors selectivity towards P1
→ High temperature completely shifts equilibrium reaction from ZP2 towards P1
→ E4 temperature sensitivity critical

Reactor Design:
→ 2 Plug flow reactors with different temperature profiles (low and high)

Separation System:
→ Solid ZP2 separation avoidable, if temperature sensitive E4 is totally consumed in first reactor
→ Dividing wall column feasible (conditions need to be simulated)

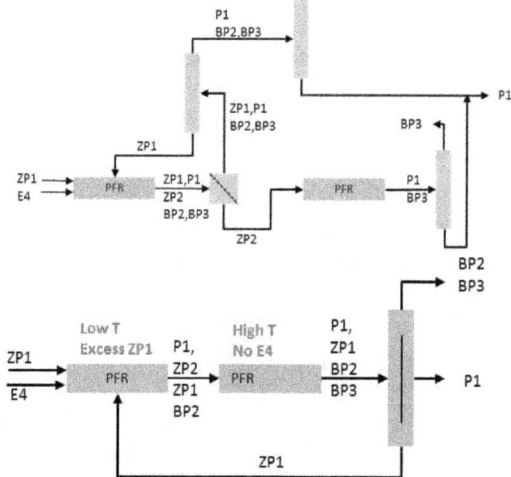

Figure 156: Sub-System 3 – Summary

recycled to the reactor system.

The liquid effluent from the solid separator contains ZP1, P1, BP2, and little BP3. Only ZP1 is recycled to the reactor to be converted into P1.

Figure 156 describes the 2-reactor system with a solid separator and three distillation columns.

Are there options for process integration? The earlier evaluation already indicated that two reactors at different temperature levels are preferable. Do we need the solid separation between reactor 1 and 2. Obviously, the reactor effluent could directly be fed to the high-temperature reactor provided E4 is completely consumed in reactor 1. Since no temperate sensitive E4 is left, there are no detrimental reactions possible in high temperature reactor 2.

The effluent of reactor 2 contains P1, ZP1, BP2, and BP3. The boiling points of these four components theoretically allow a dividing wall column to recover P1 as pure side product, while ZP1 is collected as bottom product to be recycled to the first reactor and BP3 and BP3 are removed as top products from the DWC. A separation of BP2 and BP3 is of course not necessary. Once more, it is important to emphasize that temperature, pressure and concentration levels have to be suited to operate a dividing wall column.

Design of Sub-System 4

The reaction system 4 as described in Figure 157 represents a catalytic equilibrium reaction. This system generates two small volume by-products BP_L and BP_H – low

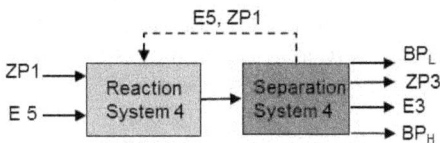

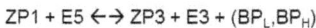

$ZP1 + E5 \longleftrightarrow ZP3 + E3 + (BP_L, BP_H)$

fast, heterogeneous catalytic equilibrium reaction,
Boiling Point P: $E3 < BP_L < E5 < ZP1 < ZP3 < BP_H$
Equilibrium Weight%: ZP3 40%, ZP1 35%, E5 14 %,
         E3 10%, BP_L, $BP_H < 1\%$

Equilibrium Reaction:
→ no issue with selectivity, maximize productivity
→ effluent with ZP1, E5, E3, ZP3, BP_L, BP_H, solid
   catalyst

Reactor Selection:
→ Inexpensive stirred tank

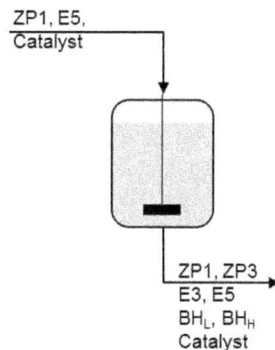

Figure 157: Sub System 4 – Reaction (Level 2)

and high boiling components (that are not part of the equilibrium reaction).

A decision on temperature levels requires more information on reaction kinetics. This, however, does not prevent the process designer from deriving a process flow sheet, since there are no competing reactions depending on temperature or concentration levels.

Figure 148 also describes the equilibrium composition for the reaction partners generally provided by development chemists. An inexpensive stirred tank reactor is once more the first choice for the reactor. The heterogeneous catalyst is added to the reactor forming a suspension. Of course, equilibrium reactions are always candidates for integrated design schemes to utilize the equilibrium feature, a sequential design as a first approach still provides insights later to be used at design levels 4 and 5.

The reactor effluent contains all reaction partners according to the equilibrium conditions. The heuristic rules covering the recycle requirements suggest a recycle of remaining feed components – common sense covered in rule 1 of the heuristics to level 3 summed up in chapter 5. The catalyst is recycled to the reactor as well (Figure 158).

By-product E3 is recycled to sub system 1, since E3 is a raw material for ZP1 characteristic for an integrated chemical site. Actually, the fact that E3 is part of the reaction system 1 may justify a reevaluation of the sub system design selected earlier. Does a joint reactor system offer new integration options for sub system 1 and 4?

Intermediate ZP3 represents the product to be recovered from the process with high quality. By-products BP_L and BP_H need to be removed from the system. ZP3 as input material for sub system 5 is preferably be recovered as a pure component, while by-products are candidates for purge streams provided economics justify loss of other valuable components (Figure 158).

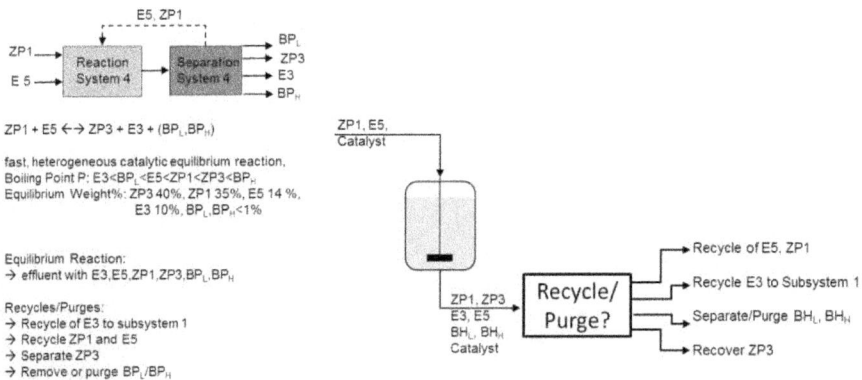

Figure 158: Sub-System 4 – Recycles/Purges (Level3)

Separation technology selections and separation sequence are next designed on level 4 of the design methodology. The selection of technologies and sequence generally may require several design loops.

A first design of the separation system does not take into account any by-products to demonstrate the problems when potential by-products are ignored or not detected during earlier process development.

Separation sequence heuristics recommend removal of solids first avoiding plugging of process equipment (level 4a: rule 3). Therefore, the suspended catalyst is separated and recycled in a first step (Figure 159). Hydro-cyclones, sieve filters, or membrane units perform solid-liquid separation depending on catalyst size.

Since all other remaining components are liquids at ambient pressure, distillation is selected as separation technology of choice (level 4a: rule 2 and level 4b: rule 1).[26]

A first distillation column is used to separate E3/E5/ZP1 from ZP3 in almost equal splits (Level 4b: Rule 4). Subsequently the mixture E3, E5, and ZP1 are separated into a top product consisting of E3 and a bottom product consisting of ZP1/E5. ZP1 and E5 are recycled to the reactor and converted into ZP3 and E3.

What happens in this process design to by-products BP_L/BP_H not discovered during process developments? Generally, low-volume by-products represent a critical challenge to every separation design. These low-volume by-products could accumulate within the process, if the separation sequence does not include an

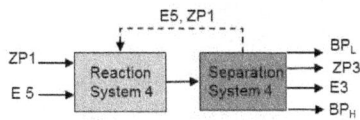

$ZP1 + E5 \leftrightarrow ZP3 + E3 + (BP_L, BP_H)$

fast, heterogeneous catalytic equilibrium reaction,
BP: $E3 < BP_L < E5 < ZP1 < ZP3 < BP_H$
$BP_H + BP_L$ only little amounts
Equilibrium Weight%: ZP3 40%, ZP1 35%, E5 14 %,
E3 10%, $BP_L, BP_H < 1\%$

Separation Sequence:
→ Remove solid catalyst first
→ Separate $BP_L/ZP3$ from ZP1, E5, E3,BP_L
→ Separate ZP4/BP_L
→ Separate E3/BP_L from ZP1/E5
→ Separate E3 from BP_L

Remarks:
→ ZP3 is recovered as pure top product
→ BP_L is candidate for purge (provided losses of E3 is economically acceptable)

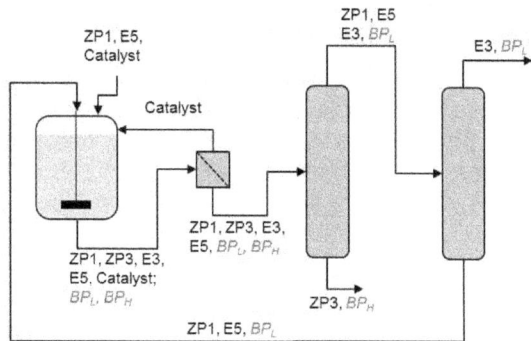

Figure 159: Sub-System 4 – Accumulation of By-products (Level3)

---

[26] Only a few exemplary heuristics are applied in this chapter.

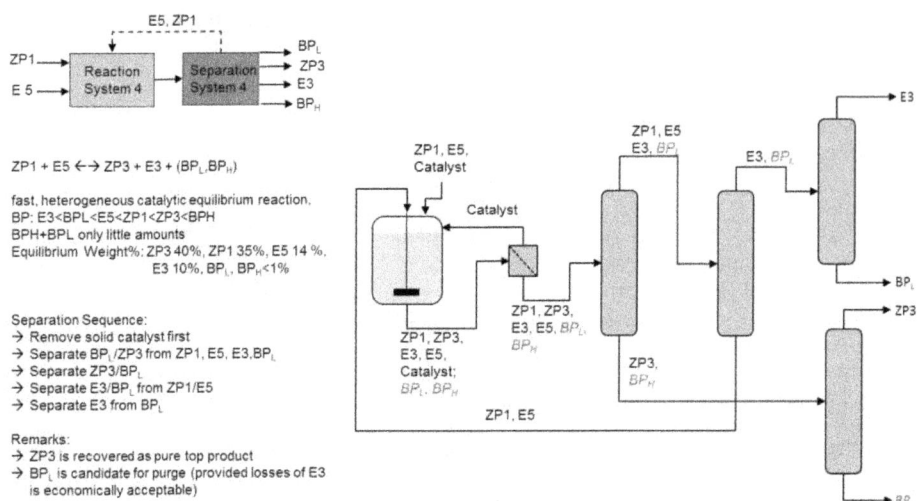

ZP1 + E5 ←→ ZP3 + E3 + (BP$_L$,BP$_H$)

fast, heterogeneous catalytic equilibrium reaction.
BP: E3<BPL<E5<ZP1<ZP3<BPH
BPH+BPL only little amounts
Equilibrium Weight%: ZP3 40%, ZP1 35%, E5 14 %,
E3 10%, BP$_L$, BP$_H$<1%

Separation Sequence:
→ Remove solid catalyst first
→ Separate BP$_L$/ZP3 from ZP1, E5, E3,BP$_L$
→ Separate ZP3/BP$_L$
→ Separate E3/BP$_L$ from ZP1/E5
→ Separate E3 from BP$_L$

Remarks:
→ ZP3 is recovered as pure top product
→ BP$_L$ is candidate for purge (provided losses of E3
   is economically acceptable)

Figure 160: Sub-System 4 – Separation Sequence (Level 4)

outlet for these by-products.

By-product BP$_H$ is always removed from the product with ZP3. Accumulation of BP$_H$ cannot occur within the system. Depending on the design of column 2, BP$_L$ may leave the system with E3 or accumulate within the system. A solution for the by-product BP$_L$ issue is to operate column 2 keeping BP$_L$ with E3 completely.

E3 and BP$_L$ are treated in another column to provide a pure E3 product reused in sub system1 as feedstock. BPL is discarded as waste. A product column finally separates the desired product ZP3 from by-product BP$_H$.

It is generally recommendable to check for the fate of small volume impurities of different boiling points generated in a process or added with raw materials. Is there a chance that these impurities may accumulate in the process? Which outlet streams may contain these impurities? Such a check is beneficial, since small quantity impurities may not be detected during laboratory or pilot testing due to limited operation time.

This sequence avoids a separation of E5 and ZP1 once more mixed in the reactor anyway. This sequence is also beneficial, since products ZP3 and E3 are recovered as distillates (level 4b: rule 7). Recovering of E3 or ZP3 as bottom product always possesses the disadvantage that a low volume, high boiling degradation by-product could affect quality of E3 and ZP3.

This sequence of Figure 160 represents the traditional process design after applying the hierarchical design procedure from level 1 (input-output structure), level 2 (reactor system), level 3 (recycle system) and level 4 (separation system). The design structure developed will now be evaluated with respect to optimization options through process intensification and integration (level 5).

Equilibrium reactions are preferred candidates for process integration to influence the equilibrium conditions favorably. A focus is given to process integration for sub system 4.

Integration aims at combining of different unit operations in one device. The reactor system uses a suspended catalyst requiring a separator to recycle the catalyst back to the reactor. An immobilized catalyst offers an alternative approach. The reaction could take place in a catalyst wall-coated tube reactor providing superior heat transfer capabilities as well. This reactor concept could be regards as an intensified and integrated reactor version, although many designers will consider this a conventional solution. Process intensification and integration actually are nothing more than a deliberate effort to search for optimization potentials.

Another standard evaluation step includes the search for integration options for reactor and separation sequence as well. Reaction columns and dividing wall columns represent the obvious approach for process integration.

A reaction column using a catalyst packing could integrate reaction, catalyst separator, and column 1 into one device. A dividing wall column could perform separations performed in two columns conventionally. Actually different dividing wall column schemes are possible to apply process integration to separations in Figure 160.

A further integration step is feasible using a reaction column with a dividing wall illustrated in Figure 161 for sub-system 5. Figure 161 includes a schematic description of component concentration profiles on both sides of the dividing wall. The catalyst is coated to the surface of the distillation packing of one side of the dividing wall internals. The materials ZP1 and E5 are fed to the column section

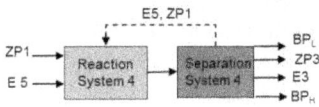

$ZP1 + E5 \leftrightarrow ZP3 + E3 + (BPL,BPH)$

fast, heterogeneous catalytic equilibrium reaction,
BP: E3<BPL<E5<ZP1<ZP3<BPH
BPH+BPL only little amounts
Equilibrium Weight%: ZP3 40%, ZP1 35%, E5 14 %,
     E3 10%, BPL,BPH<1%

Process Intensification::
→ Immobilize hetereogenuous catalyst
→ Try tube reactor with catalyst-coated walls

Process Integration:
→ Combine reaction and separation
→ Evaluate dividing wall reaction column

Remarks:
→ Dividing wall column requires suited pressure/
   temperature levels and component distribution in
   column!
→ Integrated process is more complex to operate as
   sequential approach

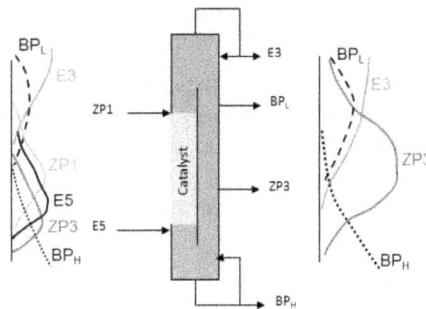

Figure 161: Sub-System 4 – Reaction Column (Level 5)

according to their boiling points above and below the catalyst section. E3 moves up the column, while ZP1 moves down the column due to their boiling points.

In the reaction section, a conversion towards ZP3 and E3 occurs according to the equilibrium conditions. ZP3 and E3 are continuously removed from the reaction section shifting the equilibrium positively towards steady production of ZP3. The feed component E5 and ZP1 steadily flow back to the reaction section within the column. Components E3 and BP_L move up the column on the left of the dividing wall, while product ZP3 and BP_H are drawn to the lower column section.

On the right of the dividing wall column product ZP3 is separated from by-product BP_H. The product ZP3 is recovered from the column at a lower section behind the dividing wall. Similarly, by-product BP_L is separated from E3 in the upper part of the column.

Of course, temperatures, pressure and concentration profiles of the different components have to allow such a design scheme. Even if this is not perfectly possible, a slight mix of streams could be acceptable. In Figure 161 a scenario is shown where the side withdrawal of BP_L contains some E3. A purge of BP_L from the reaction column is possible, if the concentration profile of BP_L allows this and loss of E3 is economically acceptable.[27]

Let us finally assume that an additional low-volume by-product appears during the equilibrium reaction due to product degradation. The boiling point of BP_M is between E5 and ZP1. What could happen to this by-product in a process design shown in Figure 160 and Figure 161?

Design of Sub-System 5

Finally, a solvent polymerization reaction is described in Figure 162. The polymerization system consists of a monomer, a linking component, and a solvent. The solvent is necessary to keep the viscosity during polymerization to a level allowing processing in standard equipment.

A stirred tank cascade, a tube reactor, or specially designed polymerization reactors perform the polymerization until the desired polymer length is achieved.

The reactor effluent contains polymer, monomer, solvent, linking compound and two by-products. A straightforward design consists of a stripping unit for low boiling by-products and ZP4. Product P2, monomer ZP3 and solvent are treated in an evaporator to remove solvent and monomer from the polymer as long as viscosity can be handled in an evaporator unit.

At low solvent and monomer concentrations, viscosity of the system increases significantly (depending on temperature). An extruder or kneader performs the final solvent and monomer removal to generate a pure polymer. A granulation unit directly combined with the extruder forms polymer granules at lower temperature

---

[27] *Potential economic benefits justify some integration efforts, even if an integration scheme initially appears unlikely.*

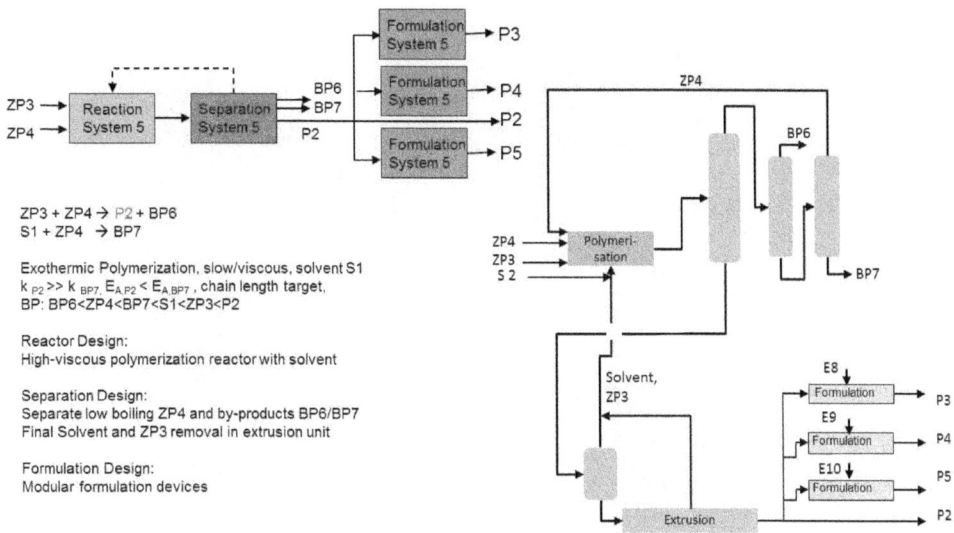

ZP3 + ZP4 → P2 + BP6
S1 + ZP4 → BP7

Exothermic Polymerization, slow/viscous, solvent S1
$k_{P2} \gg k_{BP7}$, $E_{A,P2} < E_{A,BP7}$, chain length target,
BP: BP6<ZP4<BP7<S1<ZP3<P2

Reactor Design:
High-viscous polymerization reactor with solvent

Separation Design:
Separate low boiling ZP4 and by-products BP6/BP7
Final Solvent and ZP3 removal in extrusion unit

Formulation Design:
Modular formulation devices

Figure 162: Sub-System 5 - Summary

filled in bags or silos. Solvent S2, monomer ZP3 and linking compound ZP4 are recycled to the polymerization reactor, while by-products are removed from the process.

A modular formulation unit – generally an extruder as well – is used to color or modify the granular polymer generating the tailor-made products P3, P4, P5 in addition to the base product P2. Formulation is done either at the polymer production site or at local formulation sites closer to the final customer.

Complete conceptual Design

Finally, Figure 163 shows the complete plant design. The site uses seven raw materials and two solvents. There are five processes to manufacture four intermediates ZP1, ZP2, ZP3 and ZP4 and 5 products P1, P2, P3, P4, and P5.

Reactor systems include bubble column, tube reactor, reaction column, stirred tank cascade and polymerization reactor. Solid separators, distillation columns (including dividing wall columns), extraction columns, and extrusion machines perform separations.

The individual processes are linked through material flows forming a complex process network. Process design also includes an evaluation of storage necessities within the production site to manage start-up and shutdown of the various processes or process deviations and disturbances.

A site-wide process control system is indispensable to operate the process system efficiently. This includes operation of tank farms or warehouses for raw materials, intermediates, and final products.

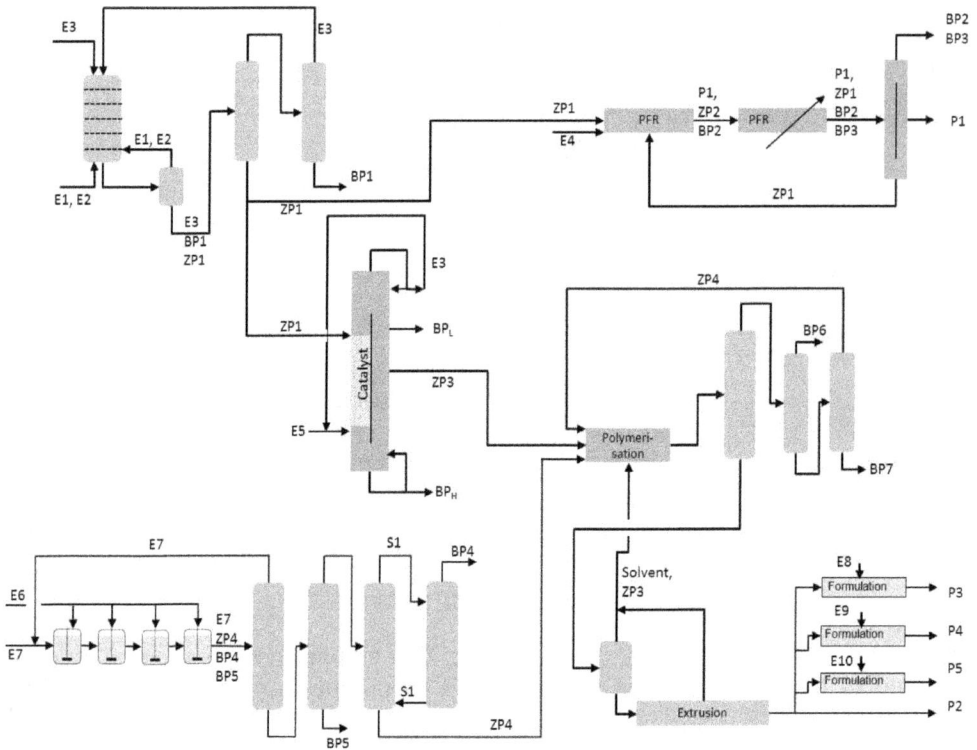

Figure 163: Complete Plant Process Design

The conceptual process and plant design as shown in Figure 163 forms the foundation for basic engineering effort. Basic engineering determines the process parameters for the conceptually designed processes verifying the feasibility of the design. In particular, cost estimates are calculated in more detail. New insights in process characteristics and cost as well may require a step back, if basic engineering discovers unacceptable process or cost performances.

Basic engineering represents the last chance to correct major flaws in the design.

4.2 Economic Evaluation

The last chapter outlined a technical design methodology. Of course, technical design must be accompanied by economic evaluations. Chapter 3.2.2 already introduced the concept of economic targets and potential to guide the technical design process.[28]

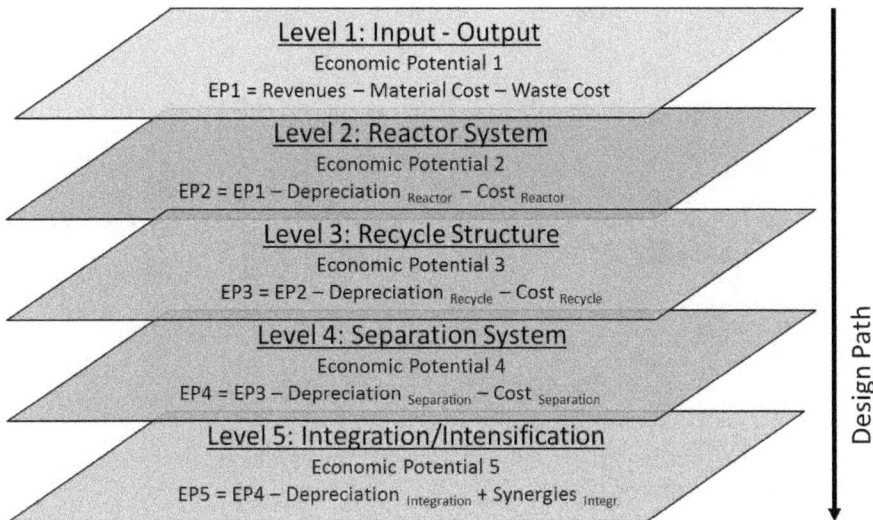

Level 1: Input - Output
Economic Potential 1
EP1 = Revenues − Material Cost − Waste Cost

Level 2: Reactor System
Economic Potential 2
EP2 = EP1 − Depreciation $_{Reactor}$ − Cost $_{Reactor}$

Level 3: Recycle Structure
Economic Potential 3
EP3 = EP2 − Depreciation $_{Recycle}$ − Cost $_{Recycle}$

Level 4: Separation System
Economic Potential 4
EP4 = EP3 − Depreciation $_{Separation}$ − Cost $_{Separation}$

Level 5: Integration/Intensification
Economic Potential 5
EP5 = EP4 − Depreciation $_{Integration}$ + Synergies $_{Integr.}$

Design Path

Figure 164: Hierarchical, Economic Potential Approach

The hierarchical economic potential concept is illustrated in Figure 164 (see comparison with hierarchical, technical methodology in Figure 147).

The process to manufacture bioethanol is used to discuss the economic potential procedure. A biotechnical process using a renewable glucose source produces bioethanol.

Figure 165 gives a schematic overview on bioethanol process using wheat as feedstock. Grain is first milled and then starch is converted into glucose by an enzymatic approach.

Microorganisms (yeast) convert glucose into ethanol during fermentation. Carbon dioxide is the main by-product. The fermentation process mainly generates ethanol, biomass, carbon dioxide, some small volume by-products, and heat. The efficiency of the microorganism defines how much glucose is converted into

---

[28] *While technical design methodologies generally follow the design procedure strictly, the economic evaluation varies with the design task.*

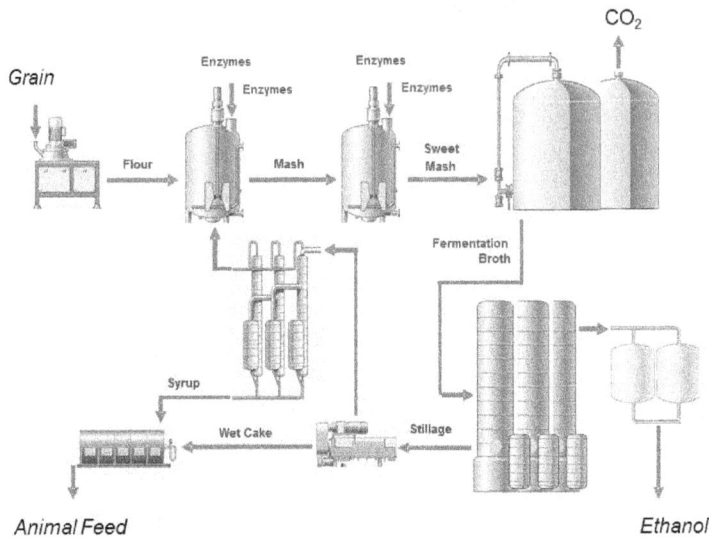

Figure 165: Bioethanol Process

ethanol and what portion is actually required to generate biomass and drive the conversion process energetically.

Downstream processing consists of ethanol dehydration, waste treatment, and biomass drying. In addition to ethanol, the dried biomass is sold as feed for animals.

Material and product prices, investment cost, and yields are only used for demonstration purposes significantly deviating from real world figures. A detailed economic evaluation is only possible after engineers have developed a final design. During the design phase of new processes, however, there is only fragmented information on the process structures available. A hierarchical approach is used to overcome that dilemma. The economic model is refined, while the design progresses.[29]

Economic potentials characterize the profitability potentials during process design. Obviously the economic potential decreases from step to step, when cost information become broader and more detailed.

Economic potential EP1 takes only material, waste and product cost into account (level 1 in Figure 164). EP1 corresponds to the first level of the hierarchical technical design approach. A major focus of EP1 is given to the impact of process yield, raw material cost and product price variation on profitability. Since neither depreciation nor operating expenses are considered in economic potential EP1, a

---

[29] Once more, the data describe a fictional case, but reflect the main issues properly.

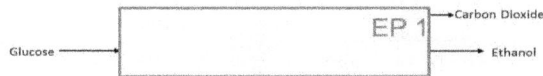

- Process:
  - Glucose → 2 Ethanol + 2 Carbon Dioxid
  - 180g/mol   46g/mol   44g/mol
  - Actual reactor yield C: 46 kg Ethanol/kg Glucose Actual recovery yield: 90%
- Prices & Volumes:
  - Glucose: 0.4 €/kg Ethanol: 1.2 €/kg Capacity: 100 000 tons/a Ethanol

- Economic Potential EP1:
  - Maximum process yield (stoichiometric, no recovery losses)?
    - 0.511kg Ethanol/kg Glucose = 100%
  - Actual reaction yield (before recovery)?
    - 0.46 kg Ethanol/0.511 kg Ethanol = 90%
  - Actual total process yield?
    - 81% → 0.414 kg Ethanol / kg Glucose
  - Maximum economic potential 1 (Stoichiometric reactor yield, no recovery losses)?
    - Ethanol Earnings: 100000 tons/year * 1200 €/ton = 120 Mio. €/year
    - Glucose Cost: 100000/ 0.511 → 195695 tons/year * 400 €/ton = 78 Mio. €/year
    - Maximal Economic Potential EP1:   42 Mio.€ year
  - Actual economic potential?
    - Ethanol Earnings: 100000 tons/year * 1200 €/ton = 120 Mio. €/year
    - Glucose Cost: 100000/0.414 → 241546 tons/year * 400 €/ton = 97 Mio. €/year
    - Actual Economic Potential EP1:   23 Mio.€ year

Figure 166: Economic Potential 1

negative or even small positive EP1 prohibits a process design, at all – at least different synthesis pathways or alternate raw materials need to be found.

Economic potential EP1 already provides an experienced process designer some indications whether this design effort will ever result in a profitable design.

Figure 166 illustrates the economic potential EP1 for the bioethanol process. Obviously, raw material cost and bioethanol prices are decisive for the economic potentials. Raw material availability and political regulations affecting feedstock and product prices can affect overall economic performance dramatically as characteristic for commodities (and actually reflected in the recent history of bioethanol and biodiesel). Businesses with former high profitability suddenly become an economic disaster.

Therefore, design must aim at maximum yield providing a sufficient economic potential at level 1. In this demonstration case, the difference between maximum EP1 (no losses at all) and the EP1 based on actual data accounts for 19Mio. €/year. An increase of the overall yield from 81% to 85% will increase the EP1 by 5 Mio. € - almost 20%. Similarly, the impact of falling product or feedstock prices on the economic potential could be estimated to develop a feeling on the volatility of the process design.

Economic potential EP2 adds the investment (better depreciation) and the operating expenses of the reaction system to the economic evaluation (EP2 = EP1 – Reactor System Depreciation – Reaction System Operational Cost).

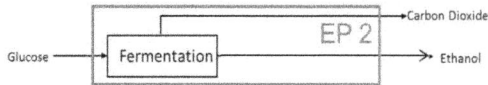

- Fermentation Systems:
  - Glucose → 2 Ethanol + 2 Carbon Dioxide
  - Capacity: 100 000 tons/a Ethanol
  - Fermenter Investment (Fermenter, aeration, feed/harvest tanks, pumps, building etc.):   30 Mio. €
  - Depreciation period: 10 years
  - Annual operating cost for fermentation (energy, labor, lease etc. / without raw materials):   6 Mio. €/a

- Economic Potential 2:

    EP2 = EP1 – Depreciation Reactor – Cost Reactor = +(23-3-6)=14 Mio. €/a

Figure 167: Economic Potential 2

Fermentation represents the reaction system in a biotechnical process. Equipment (fermenters, blowers, raw material milling/saccharification, pumps, piping, buildings etc.) define the capital investment for the fermentation system. Operational cost includes labor, energy, etc. to run this system.

Figure 167 summarizes a simplified economic potential EP2 calculation. In this case, the reaction system (level 2) adds annual cost of 9Mio. €/year reducing the remaining potential EP2 to 14 Mio. €/year. EP2 describes the change in profitability of the process with design progress. More important, it could be used to compare alternate fermentation concepts.

Oxygen needed by microorganisms can be supplied by air. Fermentation intensification is possible, if higher pressure and/or pure oxygen is used instead of air at lower pressures. Pure oxygen and higher pressure affect partial oxygen pressure in the aeration bubbles and increase oxygen solubility in the fermentation broth. High-pressure fermenters are more expensive than standard pressure vessels. Pure oxygen produced from air with pressure swing adsorption units is significantly more expensive than air. Fermentation productivity increases with pure oxygen and higher pressure consequently reducing fermenter size and capital investment – a classical optimization topic.

Simple EP2 estimates confirm that capital investment is very unlikely to justify a process-intensified fermentation for commodity products.

Economic Potential EP3 addresses the recycle economics. What are the economics of material recycles? EP3 adds the investment and operational cost of the recycle operation to the economic potential consideration (EP3 = EP2 – Recycle Depreciation – Operational Recycle Expenses).

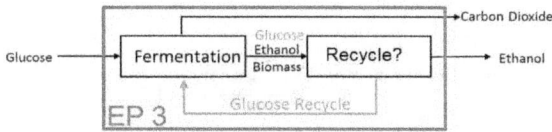

- **Alternatives:**
  - Alternative 1: Fermenter Cascade to convert glucose completely, no recycle
  - Alternative 2: Recycle of glucose with smaller fermenter sizes
    - Additional Investment for separator/recycle piping and pumps: 7 Mio.€
    - Operational Cost: 1 Mio. €/a
    - Fermenter Investment Reduction: -2Mio.€
    - Fermentation Operational Cost: -0.5 Mio.€/a
- **Economic Potential EP 3:**
  - Alternative 1:
    - EP3=EP2=14 Mio.€/a
  - Alternative 2:
    - EP3=EP2+(-0.7+1-0.2-0,5) Mio.€/a=13 Mio. €/a

<u>Figure 168: Economic Potential 3</u>

Fermenter effluent consists of ethanol, unreacted glucose, biomass, and by-products. Air and carbon dioxide is actually removed directly from the fermenter top. There are two alternative with respect to glucose – fermentation is run either until glucose is completely consumed or glucose remaining in the fermenter effluent is recycled. Economic potential 1 already emphasized the significant impact of glucose conversion on profitability making minimal glucose losses a main design target.

Figure 168 compares both approaches. If fermentation is performed until glucose is completely consumed, a recycle of glucose is not necessary. EP3 is identical to EP2 without recycle.

Once more fermentation productivity is improved, if the fermentation must not be run to completion (no glucose left), since consumption rate of the microorganisms drops at low glucose levels.

Glucose recycle decreases fermenter investment, while additional capital for recycle equipment (glucose recovery, pumps and piping) is necessary. Operational expenses for fermentation are also reduced, while operational expenses for the recycle operation are added. In addition, a recycle creates sterility challenges.

Figure 168 indicates that economics do not favor recycle operation for bioethanol plants.

Economic Potential EP4 regards the separation system. Depreciation and operational expenses of the separation system are subtracted from EP3 to get the EP4. EP4 is used to compare various separation technologies and separation sequences as well. Figure 169 illustrates the impact of the purification/dewatering system of a bioethanol process on the economic potential.

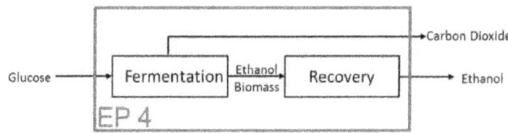

- Separation Systems:
  - Glucose → 2 Ethanol + 2 Carbon Dioxide
  - Capacity: 100 000 tons/a Ethanol
  - Separation Investment (Biomass removal, ethanol purification,
         waste treatment, tanks, pumps, building etc.):   40 Mio. €
  - Depreciation period: 10 years
  - Annual operating cost for separation (energy, labor, lease etc. / without raw
    materials): 4 Mio. €/a

- Economic Potential EP4:
  - EP4 = EP3 − Depreciation $_{Separation}$ − Cost $_{Separation}$ = ÷(14-4-4)= 6 Mio. €/a

Figure 169: Economic Potential 4

The conceptual design of level 4 evaluates a "fully operable" process for the first time. The economic potential 4 gives a first overall comparison of alternate process performances, particularly in case there is no opportunity for process intensification and integration.

Economic Potential 5 finally takes into account the impact of intensification or integration measures on profitability. Here savings of an intensification or integration measure is valued against the investment effort.

Heat integration is an important element for bioethanol plants, since energy demand to dewater ethanol to 99.9% is a major cost driver. In Figure 170 the

- What is the EP5 of the plant, assuming the investment cost is 30% higher than expected?
  - 8.5 → 6.35 Mio. €/a
- At what raw material cost is the EP5=0 assuming all other data are unchanged?
  - 400 → 435€/ton glucose (e.g. 8.5 Mio. → 0€)
- How much is the plant EBIT affected, if overall yield could be improved to 90%?
  - 81.1 → 85% results in EP5 from 8.5 →13.5 Mio.€/a
- Would you approve a development project for 3,6 Mio € expecting a yield improvement to 90%?
  - Yes, but....
- How much money could you additionally invest to modify the process provided a return is requested within 3 years (at 5% interest)? 10 Mio. €

Figure 170: Economic Potential Questions

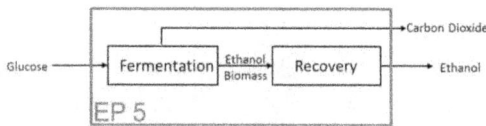

- Intensification/Integration:
  - Glucose → 2 Ethanol + 2 Carbon Dioxide
  - Capacity: $100*10^3$ tons/a Ethanol
  - Annual heat integration savings: 3 Mio. €/a
  - Investment into heat exchangers: 5 Mio. €
  - Depreciation period: 10 years

- Economic Potential 5
  - EP5 = EP4 – Depreciation $_{Int.}$ + Synergies $_{Int.}$ = +(6+3-0.5)= 8.5 Mio. €/a

Figure 171: Economic Potential 5

impact of special heat integration measures on economic potential 5 indicates savings of 3 Mio. €/year compared to an additional depreciation for heat exchangers of 2 Mio. €.

Economic potential considerations can be used to address various questions during process design.

How do uncertainties concerning investment affect profitability? Can natural raw material cost fluctuations endanger the profitability of a bioethanol plant? Is it worthwhile to invest further in process development to improve yields? Which process section offers the largest potential for economic improvements? How much money can be justified to sponsor a development project?

Figure 171 summarizes some profitability considerations.

Obviously, the economic potential concept is no rocket science. It is, therefore, particularly surprising, that these elementary considerations are often neglected in advance of starting a development project or even an investment.

The hierarchical design approach has to apply the technical and economic steps simultaneously to result in an efficient and economic design solution.

The economic potential approach – valid for the technical sibling as well – is no a rigid method, but needs to be adjusted to the individual design challenge.

5 Process Design Summary

This course on process design is a textbook that focuses on concepts, strategies, and tools. It is more about process design philosophies than detailed design calculations.

A holistic approach to process design is an indispensable prerequisite for efficient, economic, and sustainable process design. The selection of an initial process structure plays a key role for successful designs, since even an excellent engineering will not result in superior solutions, if the designer had chosen an inferior process structure.

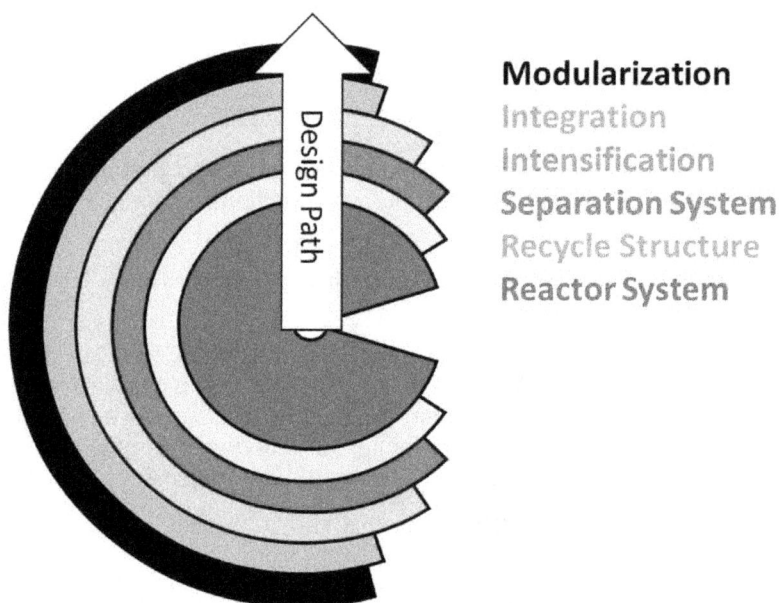

Modularization
Integration
Intensification
Separation System
Recycle Structure
Reactor System

Figure 172: Process Design Onion

Chapter 2 illustrates the basic concepts of process design concentrating on process synthesis (Leimkühler, 2010).

Figure 172 shows an "onion" description of the process design method.

Process design starts with the reactor system followed by the decision on recycles and purges. The process designer continues with the selection of separation technologies, the definition of the separation sequences, and the design of the individual separation units.

These three "inner" core steps lead to an initial process structure with preliminarily optimized parameters.

Process intensification and integration further improve this initial process flow sheet with respect to unit operation efficiency and system synergies.

Finally, the designer chooses the best capacity expansion strategy that allows a flexible, robust operation of the new process in a volatile future.

Figure 173 gives a schematic description of the process design methodology. Chapter 3 covers the main six steps of a process design. A process designer generally repeats these steps in a trial and error approach several times, before he will reach the final process design.

Chapter 6 gives an overview on heuristic rules for process design. These rules guide the designer, but do not guarantee the optimal design solution.

The process design methodology could differ in details depending whether a grass root, a retrofit, or an optimization project is the ultimate goal. Nevertheless, the conceptual aspects do not change at all.

Since this book focuses on the design philosophy, this textbook tries to communicate methodological strategies and a proper mind-set. This textbook,

Process Design Methodology:

Reactor System (1)

Recycle (2)

Separation (3)

Intensification (4)

Integration (5)

Modularization (6)

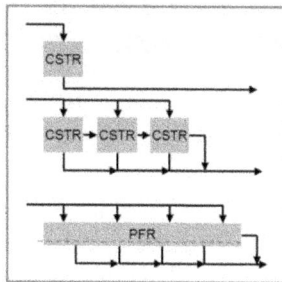

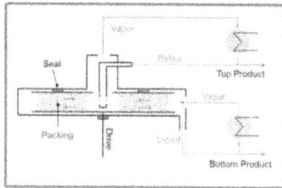

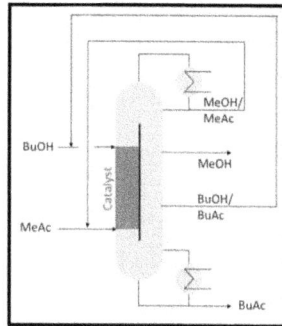

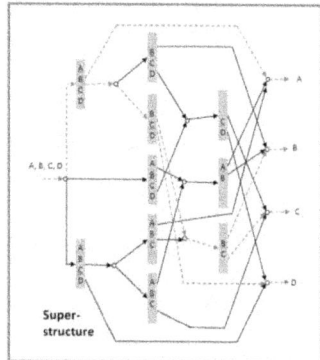

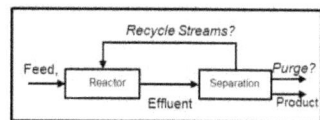

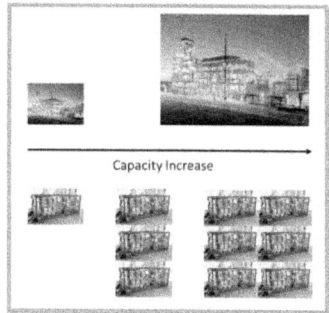

Figure 173: A schematic Design Description

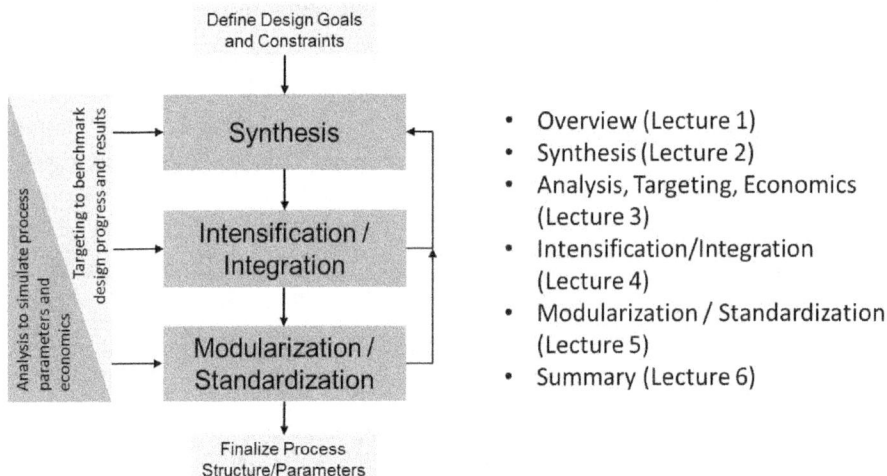

Figure 174: Course Structure

therefore, concentrates on conceptual aspects. Many chemical engineering textbooks available provide many design details and treat the various topics excellently.

The references listed in chapter 7 are books that cover different topics (such as reactor design, process intensification, etc.), but still keep a broad conceptual approach to process design in mind. The reference list does not include any reference to specialty books or scientific papers, since they will not add to the basic understanding of chemical engineering students for process design.[30]

The final Figure 174 summarizes the various topics of the course "Process Design" that I have given at ECUST and RUB indicating that the art and science of process design is an exciting topic.

---

[30] *Students are of course encouraged to use scientific books, journals, and papers in the library and more likely the internet to develop their skills in process design – in particular if process design becomes their professional field.*

6 Appendix: Heuristic Rules

This is a summary of heuristic rules to develop a conceptual design of a new process. It is important to emphasize that these heuristic rules are only a <u>limited</u> collections. There are many more heuristics available.

Heuristic rules recommend actions based on experience. They give orientation during process design, but may also lead to bad solutions due to their simplified character. Heuristics by definition are sometimes <u>ambiguous and contradictory</u>. Different heuristics could result in conflicting results. Process designers should always <u>question the results</u> and <u>ask experts for additional advice</u>. Process design is a team effort - to learn from others is indispensable (and rewarding).

<u>Do use heuristics cautiously.</u>

Methodology Rules:

Rule 1: If "several rules applicable", then "apply rule first found"

Rule 2: If "several rules applicable", then "apply rules in different sequences and check results for sensitivity"

Rule 3: If "rules are ambiguous", then "apply first prohibitive rules"

Level 1: Input – Output Rules:

Rule 1: If "facts do not obviously favor a batch, multi-purpose plant", then "start with a continuous, dedicated process design"

Rule 2: If "tailor-made technologies promise better yield and productivity", then "design dedicated processes"

Rule 3: If "product volumes are large and demands are constant", then "use dedicated plant"

Rule 4: If "chemical site or processes allow more efficient, integrated operation", then "design dedicated processes"

Rule 5: If "sufficient utilization of a multi-product plant is not possible", then "use dedicated processes"

Rule 6: If "products can be processed in identical equipment", then "consider multi-purpose plant"

Rule 7: If "unit operations occur at comparable temperature and pressure level", then "consider multi-product plant"

Rule 8: If "product volumes are small, demand fluctuating and seasonal", then "try multi-purpose plant"

Rule 9: If "batch vs. continuous decision is difficult", then "try continuous first"

Rule 10: If "batch is not obviously better", then "start with a continuous design"

Rule 11: If "start-up/shut-down creates problems", then "prefer continuous"

Rule 12: If "earlier decision led to a multi-product plant", then "consider batch operation"

Rule 13: If "earlier decision led dedicated processes", then "consider continuous operation"

Rule 14: If "later reactor and/or separator synthesis show batch operation superior", then "switch from continuous to batch mode"

Rule 15: If "synergies between processing steps cannot be excluded", then "include processing steps in one sub systems"

Rule 16: If "reactions or separations occur at similar temperature and pressure levels", then "try single reactor or separation system in a sub system"

Rule 17: If "components of different reaction systems do not form undesired by-products", then "try single reactor system"

Rule 18: If "components of different separation systems do not form undesired by-products", then "try single separation system"

Rule 19: If "reaction system conditions are different, but components do not react with each other", then "use 2 reaction systems, but only one separation system"

Rule 20: If "educts or intermediates can be purchased or sold separately", then "use separate reaction and separation sub systems"

Level 2: Reactor Design Rules:

Rule 1: A + B → C (not parallel or consecutive)

If you "have a non-competing reaction system", then "maximize rate and therefore minimize reactor size"

Rule 2: Parallel, consecutive or equilibrium reactions

If it "is a parallel or an equilibrium reaction", then "maximize selectivity and therefore minimize raw material usage"

Rule 3: Economics

If "raw material is expensive, product separation difficult and equipment cost low", then "optimize reactor design for selectivity"

Rule 4: Temperature

If "activation energy E1 of product1 is higher than E2 for product2", then "higher reaction temperature favors reaction (rate, selectivity) towards product 1"

If "activation energy E1 is equal to E2", then "higher temperature has no impact on selectivity, but favors rates"

Rule 5: Reactant Concentration

If "concentration exponent (reaction order) of desired reaction is higher than competing reaction", then "higher concentration favors desired reaction"

If "concentration exponent (reaction order) of all reactions is equal", then "higher concentrations do not affect selectivity, but increase rate of reaction"

Rule 6: Consecutive Reactions: $AN \rightarrow R \rightarrow S$

If "you want to maximize concentration of intermediate R", then "avoid back-mixing (don't mix different A to R concentrations)"

Rule 7: Economics

If "optimal selectivity requires low temperature and concentrations", then "this leads infinitely large reactors, therefore an optimum needs to be chosen"

Rule 8: Parallel, consecutive Reactions: $A+B \rightarrow R$ and $R+B \rightarrow S$

If "parallel, consecutive reactions included", then "avoid back-mixing of product R"

Rule 9: Parallel, consecutive Reactions: $A+B \rightarrow C$ and $A+B \rightarrow D$

$$r_C = k_1\, e^{-E_1/RT}\, c_A{}^{a1}\, c_B{}^{b1} \text{ and } r_D = k_2\, e^{-E_2/RT}\, c_A{}^{a2}\, c_B{}^{b2}$$

If "$a_1>a_2$ and $b_1<b_2$", use "CSTR"   (both c_A and c_B low)

If "$a_1>a_2$ and $b_1>b_2$", use "Semi Plug Flow"   (low c_A and c_B high/add A))

If $a_1>a_2$ and $b_1<b_2$", use "Semi Plug Flow"   (both c_A and c_B low/add B)

If "$a_1>a_2$ and $b_1>b_2$", use "Plug Flow"   (both c_A and c_B high)

Rule 10: Parallel, consecutive Reactions: $A \rightarrow C$ and $A \rightarrow D$

$$C \text{ desired with } r_C = k_1\, e^{-E_1/RT}\, c_A{}^{a1} \text{ and } r_D = k_2\, e^{-E_2/RT}\, c_A{}^{a2}$$

If "$a_1<a_2$", use "CSTR"   (c_A low)

If "$a_1>a_2$", use "Plug Flow"   (c_A high)

Rule 11: Parallel Reaction

If "product could react with feed to form by-products", then "avoid back-mixing and use product removal (if possible)"

Rule 12:

If "lower concentration of a component is favorable for selectivity", then "use reactor with gradual feed addition or product removal options"

Rule 13:

If "temperature profile is affecting selectivity", then "use reactor with locally adjustable temperature profiles"

Rule 14: Very fast Reactions

If "reaction is very fast", then "reactant mixing is important and reactor with efficient mixers, mixing elements is required"

Rule 15: Equilibrium Reactions

If "you deal with an equilibrium reaction", then "try to immediately remove product from reactor"

Level 3: Recycle Design Rules:

Rule 1: If "feed materials are not completely reacted", then "recycle feed components to reactor inlet after separation from product"

Rule 2: If "reactor effluent contains solvents", then "recycle solvent to reactor"

Rule 3: If "by-products occur in equilibrium reaction", then "recycle them to extinction"

Rule 4: If "inert impurities and irreversible by-products are generated", then "separate them from product and do not recycle them"

Rule 5: If "Inert impurities and irreversible by-product streams contain only little feed, solvent or product", then "evaluate impurity purge stream"

Rule 6: If "small volume by-products or impurities are possible", then "check system for accumulation potential"

Level 4a: Separation Design Rules:

Rule 1: If "alternate separations feasible", then "look for separation technologies with large driving forces"

Rule 2: If "components liquids at standard temperature", then "use distillation"

Rule 3: If "mixtures contain solids", then "remove solids from liquids, vapors, or gases first"

Rule 4: If "selecting separation technologies", then "prefer technologies with only liquid phases"

Rule 5: If "selecting separation technologies", then "try to avoid additional solvents"

Rule 6: If "selecting separation technologies", then "avoid extreme conditions"

Rule 7: If "separation technologies could damage product", then "try to avoid them"

Rule 8: If "product is a commodity", then "try low cost separation technologies"

Rule 9: If "molecules large, sensitive, or biological", then "use membrane techniques or chromatography (adsorption)"

Rule 10: If "solid products desirable", then "use crystallization for purification"

Rule 11: If "products are heat-sensitive", then "do not use distillation type separation systems"

Rule 12: If "products are heat sensitive", then "use of extraction, membrane technologies, crystallization, and particularly chromatographic methods"

Rule 13: If "products have very close boiling points", then "do not use distillation type separations systems"

Rule 14: If "alternatives to solvents are available", then "do not use solvents requiring additional separation and recycle equipment"

Rule 15: If "products differ in molecular size", then "try to use membrane technologies or chromatography (main disadvantage is fouling and membrane stability)"

Rule 16: If "products have different solubility in solvent", then "evaluate use of crystallization requiring a solid separation process and mother liquor treatment"

Level 4b: Distillation Sequence Rules

Rule 1: If "all components liquid at ambient temperature", then "use distillation type separation"

Rule 2: If "component sensitive, corrosive, toxic, or dangerous", then "remove first"

Rule 3: If "component quantitative dominating mixture", then "remove first"

Rule 4: If "component quantitative similar", then "prefer equal splits"

Rule 5: If "components difficult to separate", then "perform difficult separation last"

Rule 6: If "component is to recover with high purity", then "perform separation last"

Rule 7: If "component is to recover with high purity", then "recover as vapor product"

Rule 8: If "no other rule objects", then "perform sequence with increasing boiling temperature"

Rule 9: If "no other rule applies", then "perform most easy separation next"

Rule 10: If "small volumes of easy to separate components available", then "try side column"

Level 5a: Heat Integration Rules Integration Rules

Rule 1: Do not transfer heat from above to below the pinch temperature

Rule 2: Integrate exothermic reactors above pinch temperature

Rule 4: Integrate endothermic reactor below pinch temperature

Rule 5: Install distillation columns above or below pinch temperature (do not use evaporator above and condenser below pinch)

Rule 6: Use heat pumps to transfer heat from below pinch temperature to temperatures above pinch

Rule 7: Install heat to power devices above or below pinch

Rule 8: Divide the problem into area above and below the pinch and do not design heat exchanger across the pinch point

Rule 9: Start to add heat exchanger at the pinch point

Rule 10: Above pinch: Try to link hot stream with cold streams when CP_{Hot} < or = CP_{Cold}

Rule 11: Below pinch: Try to link hot stream with cold streams when CP_{Hot} > or = CP_{Cold}

Rule 12: Make the heat exchangers as large as possible (to get rid of complete streams). This minimizes total number of heat exchangers necessary

Rule 13: Generally, there are more alternative heat exchanger selections further away from pinch temperature possible

Rule 14: The minimum number of heat exchangers, coolers, and heaters follows: Number $_{minimum}$ = N $_{streams\ above\ pinch}$ + N $_{streams\ below\ pinch}$ (assuming there is only one heating and cooling source)

Rule 15: If there are different heat and cooling sources available:

(N $_{streams}$ + N $_{heat\ sources}$ − 1) $_{above\ pinch}$ + (N $_{streams}$ + N $_{cooling\ sources}$ − 1) $_{below\ pinch}$

Rule 16: If design is not possible, then simply try to split streams.

Level 5b: Process Integration

Rule 1: If "process includes recycle streams", then "evaluate a device integrating reaction and separation"

Rule 2: If "reactions include an equilibrium reaction and distillation separations", then "check reaction column"

Rule 3: If "distillations in a sequence occur at similar pressures and temperatures", then "evaluate dividing wall column"

Rule 4: If "reaction and/or separation occur in viscous systems", then "try to combine in a high power machine (extruder, kneader)"

Many more heuristics for process integration are feasible.

7 References

Literature:

Anderson, C. (2013). *Makers.* Random House.

Bradshaw, C., Reay, D., & Harvey, A. (2008). *Process Intensification - Engineering for Efficiency, Sustainabilit and Flexibility.* Elsevier.

Brynjolfsson, E. (2013). *The Second Machine Age.* W.W.Norton&Company, Inc.

Douglas, J. (1988). *Conceptual Design of Chemical Processes.* McGraw-Hill.

El-Halwagi, M. (2012). *Sustainable Design through Process Integration.* Elsevier.

Händeler, E. (2011). *Kondratieffs Gedankenwelt.* Brendow.

Klemes, J. (2011). *Sustainability in the Process Industry: Integration and Optimization.* McGraw-Hill.

Kohlgrüber, K. (2008). *Co-Rotating Twin-Screw Extruders.* Carl Hanser Verlag.

Leimkühler, H.-J. (2010). *Managing CO2 Emissions in the chemical Industry.* Wiley-VCH.

Mothes, H. (2015). *No-Regret Solutions - Process Design in Times of Uncertainty and Complexity.* (in press).

Ovenspiel, O. (1972). *Chemical Reactor Engineering.* John Wiley & Sons.

Peters, M. (2004). *Plant Design and Economics for Chemical Engineers.* McGraw-Hill.

Schwartz, P. (1991). *The Art of the long View.* Doubleday.

Seider, W., Seader, J., & Lewin, D. R. (2004). *Product &Process Design Principles.* John Wiley & Sons.

Smith, R. (2004). *Chemical Process - Design and Integration.* John Wiley & Sons.

Stankiewicz, A. M. (2004). *Re-Engineering The Chemical Processing Plant.* Marcel Dekker.

Sundermann, K., Kienle, A., & Seidel-Morgenstern, A. (2005). *Integrated ChemicaL Processes.* Wiley-VCH.

www.ingramcontent.com/pod-product-compliance
Lightning Source LLC
Chambersburg PA
CBHW081504200326
41518CB00015B/2374